Material+: Creative Products

Preface

This book introduces materials used in works and presents the different fields of designers' ideas, manufacturing systems, and ways of thinking. This can be well demonstrated by these newly designed products, some of them have won international contests. It also contains serious questions on why the designers want to create these works and which material does the designers prefer to use.

To designers, except their unique ideas and conceptions about creating works, their investigations of new materials are also regarded as a basic and important stage they should experienced. All designers dream of creating a novel work from common materials. Some designers are creating "accustomed but newly felt" by-products in their unique way and attempting to make changes by using of colours, processing methods. The selection of the optimal materials is a key factor for creating beautiful designs which will be faithful to their original function and role.

These special works and the creativity of these designs will bring a new and pleasant experience to readers while they are reading this special book.

Monocomplex Design Studio

ARTPOWER

Material+: Creative Products

ARTPOWER™

Designer: Rainy Wong
Chief Editor: Rainy Wong

Address: Roomc, 9/F., Sun House, 181 Des Voeux Road
Central, Hong Kong

Tel: 852 3184 0676
Fax: 852 2543 2396
URL: www.artpower.com.cn
Email: artpower@artpower.com.cn

ISBN: 978-988-16687-5-2

Printed in China

Contents

360° LAMP / 360° LAMP EASY

Designer Pavel Sidorenko
Photographer Pavel Sidorenko

"360° LAMP" is an unique light source with even the smallest details carefully planned, each element of which has a functional meaning. It is very simple to put together and during installation you can choose between a white or mirrored external surface. Your home will be filled with smooth and calming light, and the original laser engraving lends a special charm to the room. Don't be in a hurry to dispose of the "unnecessary" ring from the middle of the lamp because it makes for a good bookmark.

"360° LAMP EASY" is an unique self-assembly lamp made from recyclable polypropylene. Material chosen for it's quality, durability and recyclability. Well thought out construction makes it easy to assemble for everybody without the need for any tools or glue. This is a contemporary pendant light fit for any space and fills your home with soft calming light. And don't be in a hurry to throw away the superfluous circle from the center of the lamp, as it can be used as a bookmark.

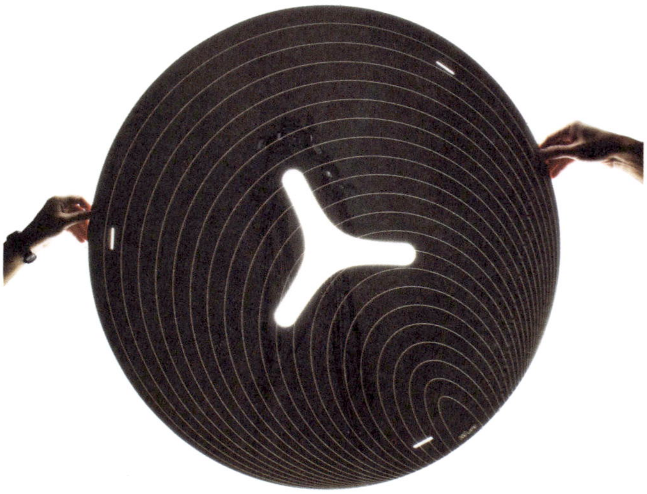

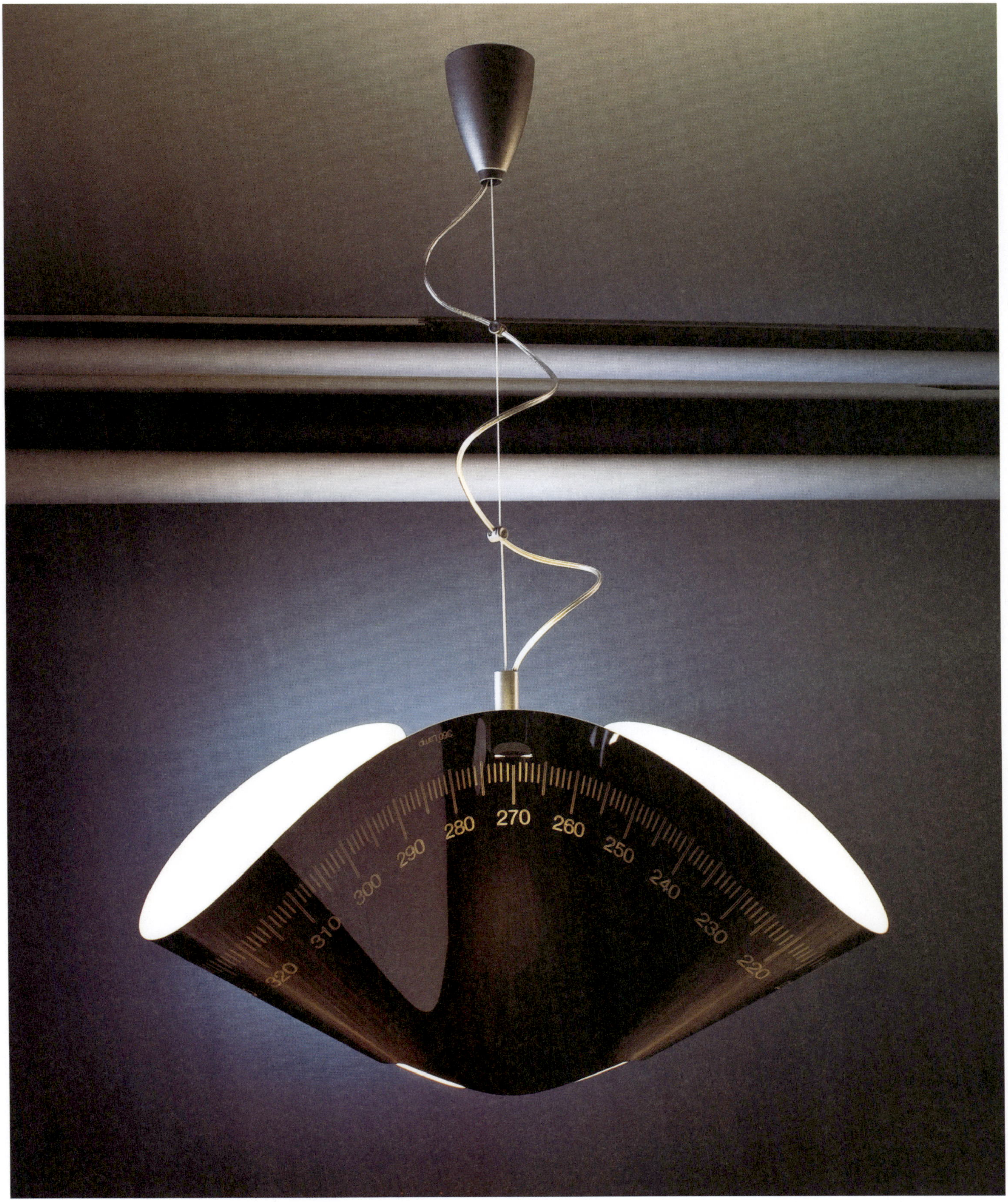
270
260
250
240
230
220
280
290
300
310
320

Free-standing Room Divider "Fence"

Designer Pavel Sidorenko
Photographer Pavel Sidorenko

Sometimes kids want their privacy, but you just don't have enough rooms to give each kid their own room. It's hard to constantly share a space with someone else, and not have an opportunity to have a little alone, or private time. Even if you only use one or two dividers and don't divide the entire room it can set some boundaries and give kids a sense of space. This is innovative cardboard room divider called fence. It is non-toxic and best of all completely recyclable. This product destined to give a lot of pleasure for your kids. The Fence was designed not only to close off a play area, but also as the object of play itself. Your children will love them. They could decorate fence and birds with their own incredible artwork! With this kids room divider children could organize a lot of games. You won't find any instruction manual, however. The only rules are that there are no rules.

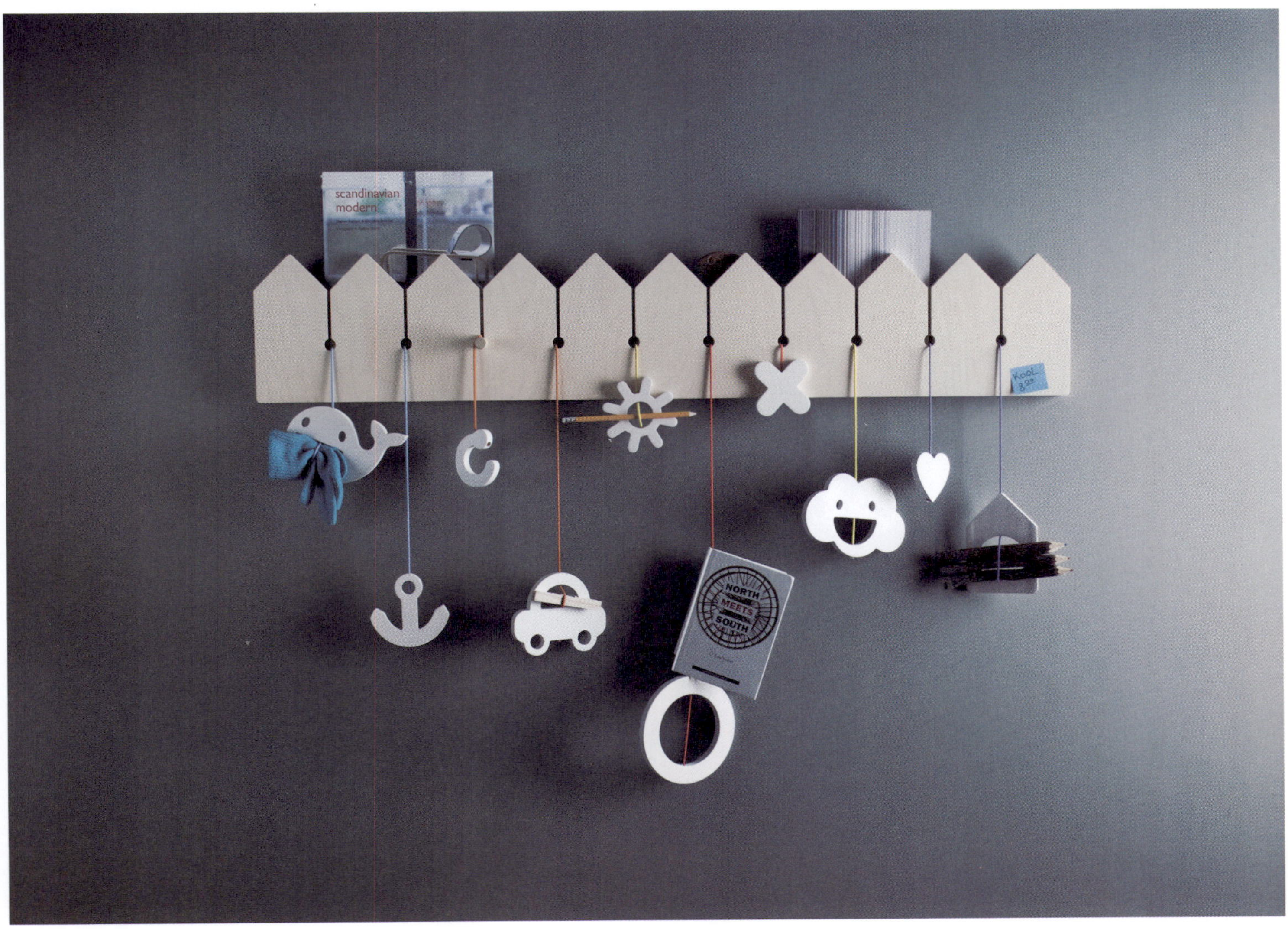
scandinavian
modern
NORTH
MEETS
SOUTH

Mustaches "Old Thomas"

Designer Pavel Sidorenko
Photographer Pavel Sidorenko

In this project is the potential for a better world. As a mustache have myself, I must say, if you haven't tried wearing one, you haven't lived. And here's your newest and best opportunity. Pick mustache from your pocket and clip them on! Wear one to the office or to a dinner party to impress your new date! Or just make people smile around you! The range has four styles to choose from and each attaches to the wearer's septum.

Cradle

Designer Pavel Sidorenko
Photographer Pavel Sidorenko

A
5
M
A
5

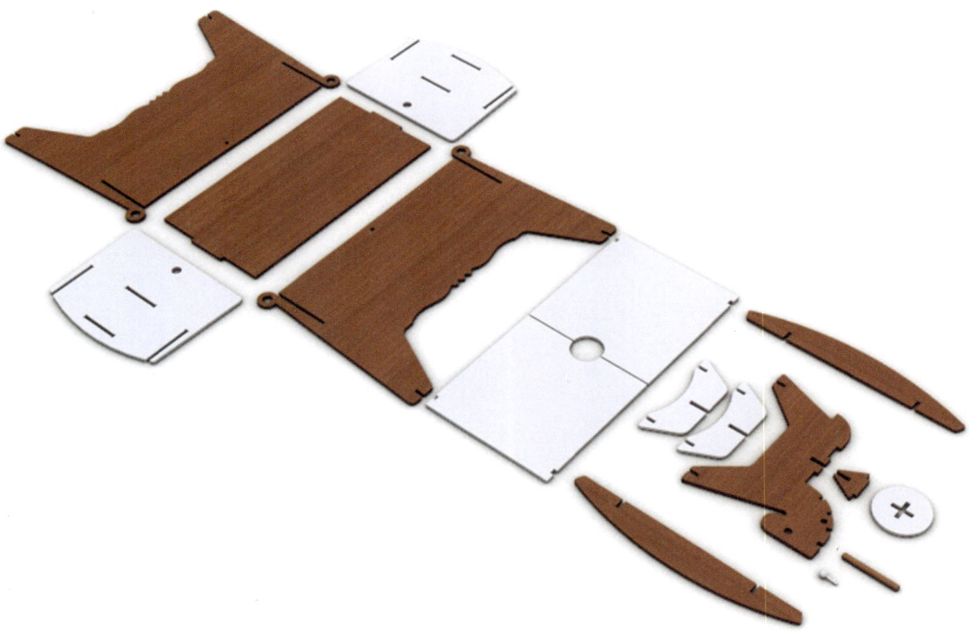

Yapu

Designer Pavel Sidorenko
Photographer Pavel Sidorenko

Re Vinyl Wall Clock

Designer Pavel Sidorenko
Photographer Pavel Sidorenko

Re Vinyl is an upcycling product fashioned from old vinyl records. By altering the functions of used products for new uses, it effectively reduces the disposal of potentially useful materials and the consumption of new raw materials at once. Discreetly crafted on the theme of scenography or animal creatures, the collection comes in a variety of story-telling silhouettes while retaining the grooves and record labels reminiscent of the heydays of acoustic recordings.

PANDION
NY
NY

EMI
INTERNATIONAL

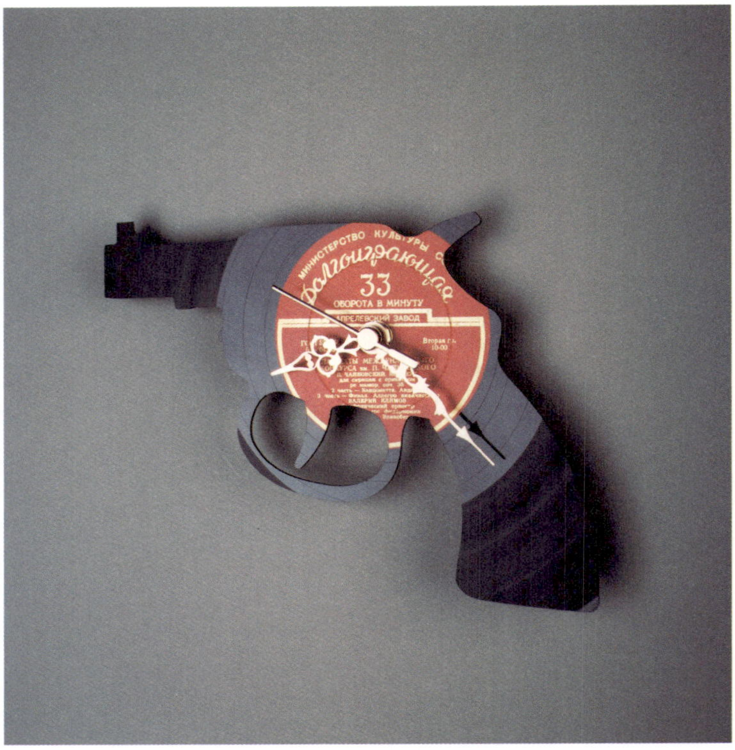
33
ОБОРОТА В МИНУТУ

БТА

МИНИСТЕРСТВО КУЛЬТУРЫ СССР
Долгоиграющая
33
ОБОРОТА В МИНУТУ
ТУ—I кл.
Б

Sectionimal

Designer gt2P
Client Self-commission
Photographer Bside Studio

The Sectionimal table belongs to the section family. It was created by digital modeling with parametric generative algorithms capable of section volumes. This table has a free volume sectioned and assembled from its manufacturing plans.

Twin Shelves

Designer gt2P
Client Self-commission
Photographer Bside Studio

The Twin Shelves is pat of the voro family, a series of furniture and objects digitally designed from an algorithm based on voronoi tessellation. An edge and points which determine the internal subdivision of the variables are identified and modified with this algorithm, which allows to customize the shape and dimensions x, y and z and the number of internal subdivisions, on the basis of the specific requirements of each client and location.

Gobbalunga

Designer Giovanni Levanti
Client Campeggi
Photographer Ezio Prandini

A traditional chair with a long, soft tail.

An unusual sofa, to relax on or for a friendly chat, whose form can vary and be continuous, broken or bent according to the function.

MOZIA

Designer	Giovanni Levanti
Client	Diamantini & Domeniconi
Photographer	Nadia Iannello

Colourful Mediterranean abstraction catches the eye.

Clock characterized by a wide, thin and cone-shaped face housing, in an unusual relationship, the minute and the hour hands. Solar synthesis turning time into an aestethic experience.

Rilassata

Designer Giovanni Levanti
Client Campeggi
Photographer Ezio Prandini

A cheerful and light small armchair for an immediate double comfort: it can change into a rug on which lying keep your legs high in order to relax completely.

Cameo

Designer	Giovanni Levanti
Client	Campeggi
Photographer	Ezio Prandini

A playing trophy instead of a hunting one.

A basket instead of a stuffed animal head with big horns.

Playing as part of everyday life. A ball scores a basket.

Home as a free pitching area.

Snack Diamond Ring–For the Love of Now

Designer Koshi Kawachi
Photographer Koshi Kawachi

The snack was decorated with the natural diamond.

Manga Farming

Designer Koshi Kawachi
Photographer Koshi Kawachi

This is an installation artwork making by planting vegetable seeds on the old comic book. The seed is planted on the artist or the farmers' favorite page as if they are the bookmarkers.

人生

泣ける話
ジャンプ
龍狼伝

「豊胸注射」で
バストアップ!!
7日以内!

◆ドカ〜ン◆
ムウ

Tea Time Packaging

Designer I.D. Monica Navarro Aranda
Client CENIVAM (Student Work)

Tea Time is an assorted tea bags package.

The brand and packaging design was based on form and structural properties of the Roselle (Hibiscus sabdariffa), a plant commonly used in teas and infusions.

Package made of printed paperboard and polypropylene sheet.

The inside is divided into five compartments covered by single layers, allowing access to each tea flavor section independently.

vanilla

better before Sep 2010
store in a cool dry place
made in Colombia
tea time
five kinds of
high quality tea
net weight
250gr

orange
water melon
green tea
vanilla
black berry

tea time
makes your day

KinderKat, Modular Playhouse for Cats

Designer I.D. Monica Navarro Aranda
Client Student Work

KinderKat is a modular furniture for domestic cats to scratch, exercise, rest and play. Designed to be fun enough to dissuade the cat from scratching other furniture or climbing curtains and become its favorite place of the house.

The user can choose the quantity of modules depending on how high he wants it to be and what he wants for his cat, such as catnip toys, sound toys, hiding place, bed or food dispenser.

Modules can be assembled at 90° or 180° to fit on corners or plane wall locations.

KinderKat's upholstery is removable and washable to assure hygiene and avoid bad smells.

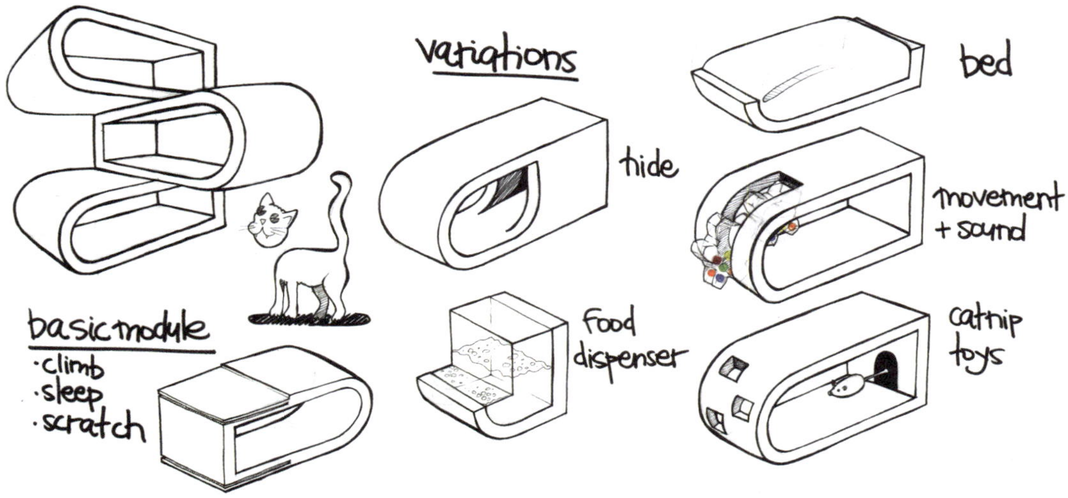
variations
bed
hide
movement + sound
basic module
·climb
·sleep
·scratch
food dispenser
catnip toys

Bright Woods Collection

Designer Giancarlo Zema Design Group
Client Avanzini Group Distributor of Quality
Photographer Giancarlo Zema Design Group

Lengthening one of its sinuous tentacles, a curious and colourful octopus emerges from the seabed, bringing shape, colour and vital light. With this vision, the architect Giancarlo Zema designs a floor lamp in DuPont™ Corian® for LumineXence. Bicoloured with a steel base, dynamic RGB LED technology with remote control, 210 cm height and 35 cm in diameter. An intriguing art-design object, in a limited edition of 120 pieces, signed and numbered.

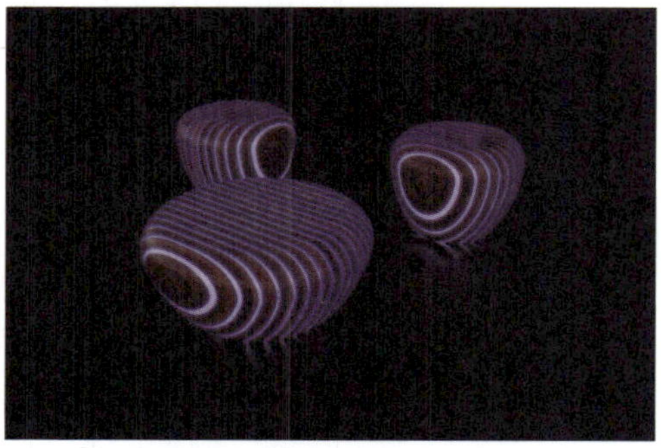

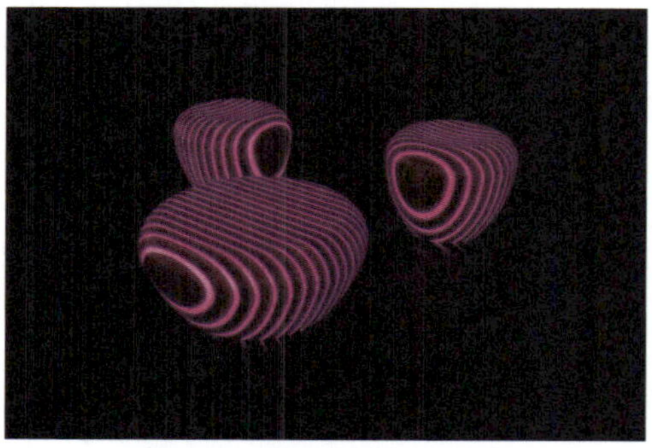

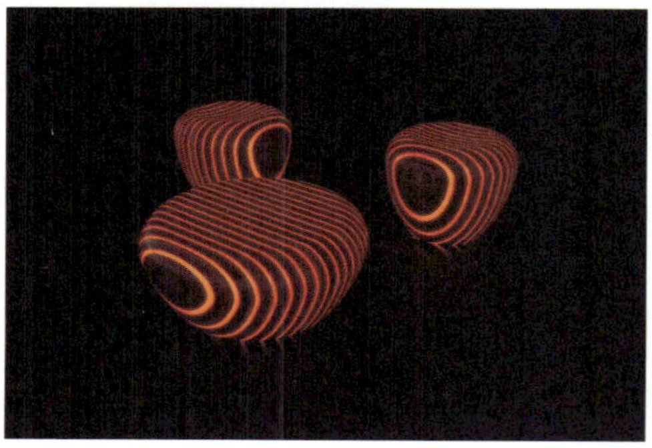

SimoGi

Designer Giancarlo Zema Design Group
Client Bsolid
Photographer Giancarlo Zema Design Group

With its oriental sounding name, SimoGi by Giancarlo Zema, joins two elements, water and earth, taking inspiration from the symbolic embracement between sea foam and an abandoned stone on the shore. A precious little object-holder in DuPont™ Corian® realized for Bsolid in limited edition of 120 signed pieces for exclusive design lovers.

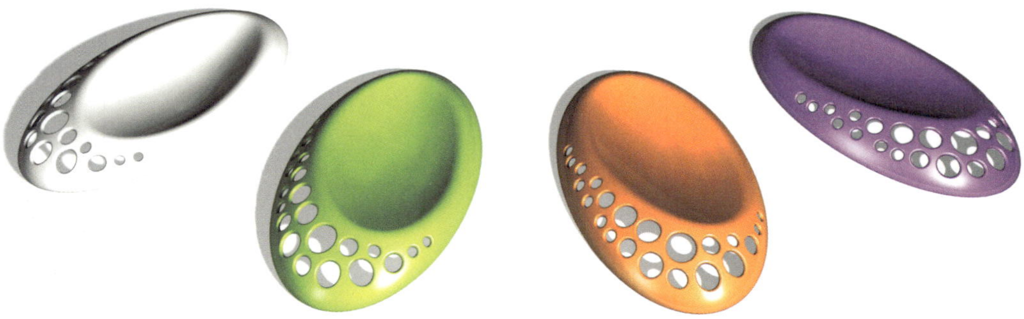

Lotus

Designer — Giancarlo Zema Design Group
Client — LumineXence Srl
Consultants — Solarday-Siderpali – Nord Zinc

Like a big leaf that comes out from the ground, Lotus gives life to an intriguing urban design shelter, endowed with seats and integrated photovoltaic panels, perfect for green parks and parking areas. A unique tubolar element of 14cm in diameter and 260cm in height, radially arranged, generates the modular strong and elegant backbone, able to cover from the rain, to light the ambient with LED technology and to produce energy: from the small leaf of 4Mq photovoltaic surface for a 500W production to the big leaf of 19Mq for 2.8KW. Lotus is also an electric car charging point: thanks to a centralized system each parking owns an info point, with a waterproof electrical outlet by Scame, from wich is possible to recharge the car, just paying with bancomat or credit card. Thanks to its endless combination of colours, offered by a wide range of strictly ecological paints, Lotus fits in perfectly with nature adapting to the colours of the enviroment; the fresh green tones for natural environments, neutral tones for urban contests, and other customizable possibilities. An innovative project by the architect Giancarlo Zema for the LumineXence from Bergamo, and sponsored by a pool of leading companies in the production of technological solution for the eco-sustainable field, like Solarday, Siderpali, Eurozeta, Nord Zinc, Scame Parre.

lumineXence

Paintbrush Mirror

Designer Monocomplex
Client Self-Project (for sale)
Photographer Monocomplex

Paintbrush mirror' is a mirror with the motif of paint rolling down from painting brush to painting the walls.

Dynamite Vase

Designer　Monocomplex
Client　Self-Project (for sale)
Photographer　Monocomplex

Dynamite invented by Nobel was a fatal weapon to destroy peace. However, the true purpose of it was to end war and bring peace other than the outbreak of war. "Dynamite Vase" is a work to give paradox and wit through a vase with the key word of love to dynamite with the key word of war and danger. "Dynamite Vase" made a glass vase held by ash wood and used leather string as finishing material to combine each of dynamites.

Dynamite Vase

Hide and Seek

Designer Monocomplex & Seunghyun Lee
Client Self Initiated Work
Photographer Monocomplex

Hide tight, I might see your hair! Hide & Seek is a furniture set made with the motive of hide and seek playing, and it is a design made with wooden material and which refreshes space by not exposing and hiding flowerpot, book, drawer handle, and etc for interior decoration, which can express more beautiful sight of interior atmosphere.

Lab Series: Vase/Single Flowervase

Designer Monocomplex
Client Self Initiated Work (For Sale)
Photographer Monocomplex

Vases are so common that we can see them anywhere. But, our interest is colorful flowers beautifully bloomed and the vases are only supporting roles. The main use of Lab Series: Vase by the approach to apply test tube as major material is a flower vase.

We attempted to study the method to have a value more than inherent beauty without damage to existing properties when a complete object is mixed with a new processed product through new consideration of existence of vase.

Retro TV Lighting

Designer Monocomplex
Client Self Initiated Work
Photographer Monocomplex

According to digital time's advent, analog things are being protruded to the back seat.

Retro TV lighting intends to deliver warmness in memories to users by moving the warm nature of wood materials along with analog emotion to the illumination.

Icicle Lighting

Designer Monocomplex
Client Self Initiated Work
Photographer Monocomplex

When winter comes, we can see the icicle hanging at the edge of eaves here and there.

The bore light at the eaves is the degree of dazzling because it is white and bright same as winter.

If we can bring the beauty to my side? Icicle attempts to give usages over illumination by adding features of the shelf to it.

Spool Lighting

Designer	Monocomplex
Client	Self Initiated Work
Photographer	Monocomplex

The spool has been used various from the past till now. Until now a spool has been used from examples so as to be Various. Specially, it is kept in mind as a symbol of representing warm figures of our mothers who are sewing.

This work makes the basic forms and usage of the spool and needle reborn to modern objet by which everyone can feel warm nostalgia after reinterpreting it through lights.

Mons Lavabo

Designer MAMEMIMOMU

Washbasin is one those elements least questioned because its conventional structure fulfills its function. Yet we think of the washbasin from the children's perspective, and that it should ideally promote hygienic behaviour. With this in mind, we forged a new vision of the child-friendly washbasin.

Save Water: With water provision is automatic, less is wasted and there is no risk of accidentally leaving the faucet open. The built-in soap dispenser also ensures a substantial saving in hand-wash agent.

Security: With the Mons form, possible accidental knocks are avoided from below. The innovative material also allows for a softer possible contact with the washbasin.

Cleanliness: The open form of the conventional washbasin makes it easy for water to splash on the floor. the closed form of Mons reduces the splashes.

For the elaboration of the Mons washbasin "Duralmond" is used, which is a material obtained from the mixture of synthetic and natural resins. Its composition is very resistant and light, waterproof, fireproof, as well as 100% biodegradable and recyclable.

Besides the functional improvements, now the child has a new, pleasant and likeable experience, which is the reason we resort to a design of attractive forms and direct association. The child introduces their hands in the smooth mouth of the figure which can be pleasant, and the fun is enhanced by the tooth element.

Mons is thought to be placed in school toilets, children parks, in other places where there are children or for the home. For that reason they are designed with different materials, colours and accessories, that would optimise the relationship between user and utensil.

mons
LAVABO

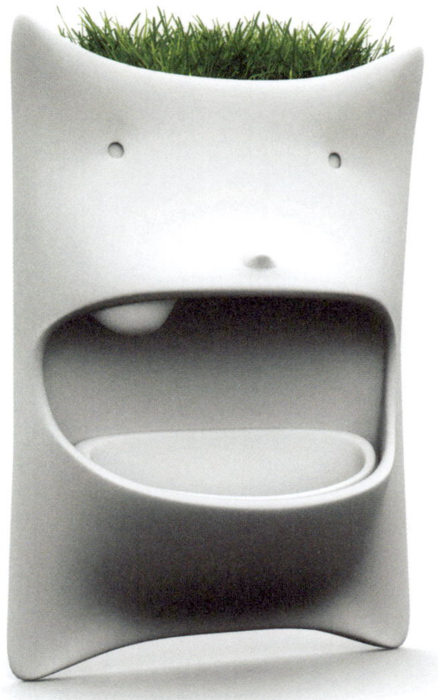

Gaufre

Designer Olga Kalugina
Client Study project

Gaufre is a support for discs. It contains in herself to 8 cd in a box. This support will serve as a good decoration for a desktop.

Bracket Clock Shade of Time

Designer Olga Kalugina
Client Work for design competition

This clock doesn't have hands. Instead of them there are shades. Around the perimeter of the clock dial there are two rows of lighttubes: the first row shows the hours and the second is for minutes. Switching on in a concrete period of time, lighttubes put their light on a small eminence in the center of the clock dial, that makes shades.

Thanks to specific features of the form, that prevent clock dial from the light, this clock can be used in daytime.

Clock-case is made from silver. The game of patches of light and reflections casts a spell and makes a special charm.

These items should remind customer that time is fleeting, it is only a shade.

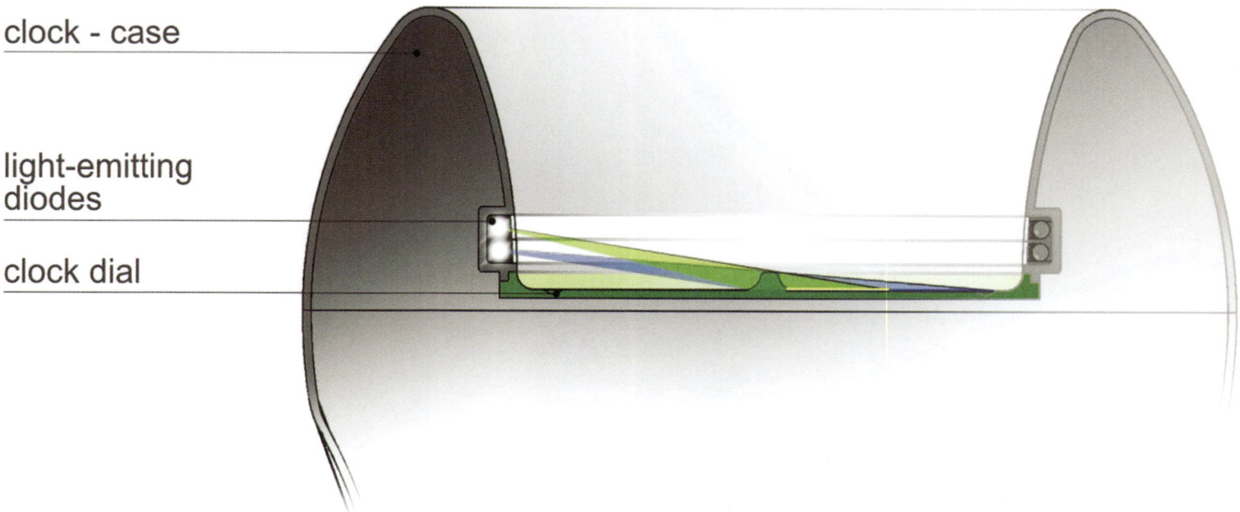

Mobile Living Module Bloom

Designer Olga Kalugina
Client Diploma project for master's degree

Mobile living module working on renewable energy sources. Mobile living module bloom is destined for rest on the nature far from a civilization. The module works completely independently on renewable energy sources. The object is equipped by solar batteries, air and liquid solar collectors, water treating filters, the furnace of slow burning and system of processing of dust.

The module can be transformed: lateral parts of the module can be turned from the central part to the outside. This increases internal space of the module. The living space includes the entrance zone, three private living zones and the general territory into which the diningzone, the bath-room and kitchen. The private living zone is some kind open room for two persons. The first berth is a top folding shelf. The second berth is transformed to a sofa and a bench in the morning, separating a workplace from the general zone.

The buyer of the module will be the travel company which will manage trips to the place chosen by the client. The module is intended for six persons from three days to one week. The module is transported by means of motor transport or the helicopter. The size of the module is multiple to the sizes of standard containers. It is possible to connect some modules together.

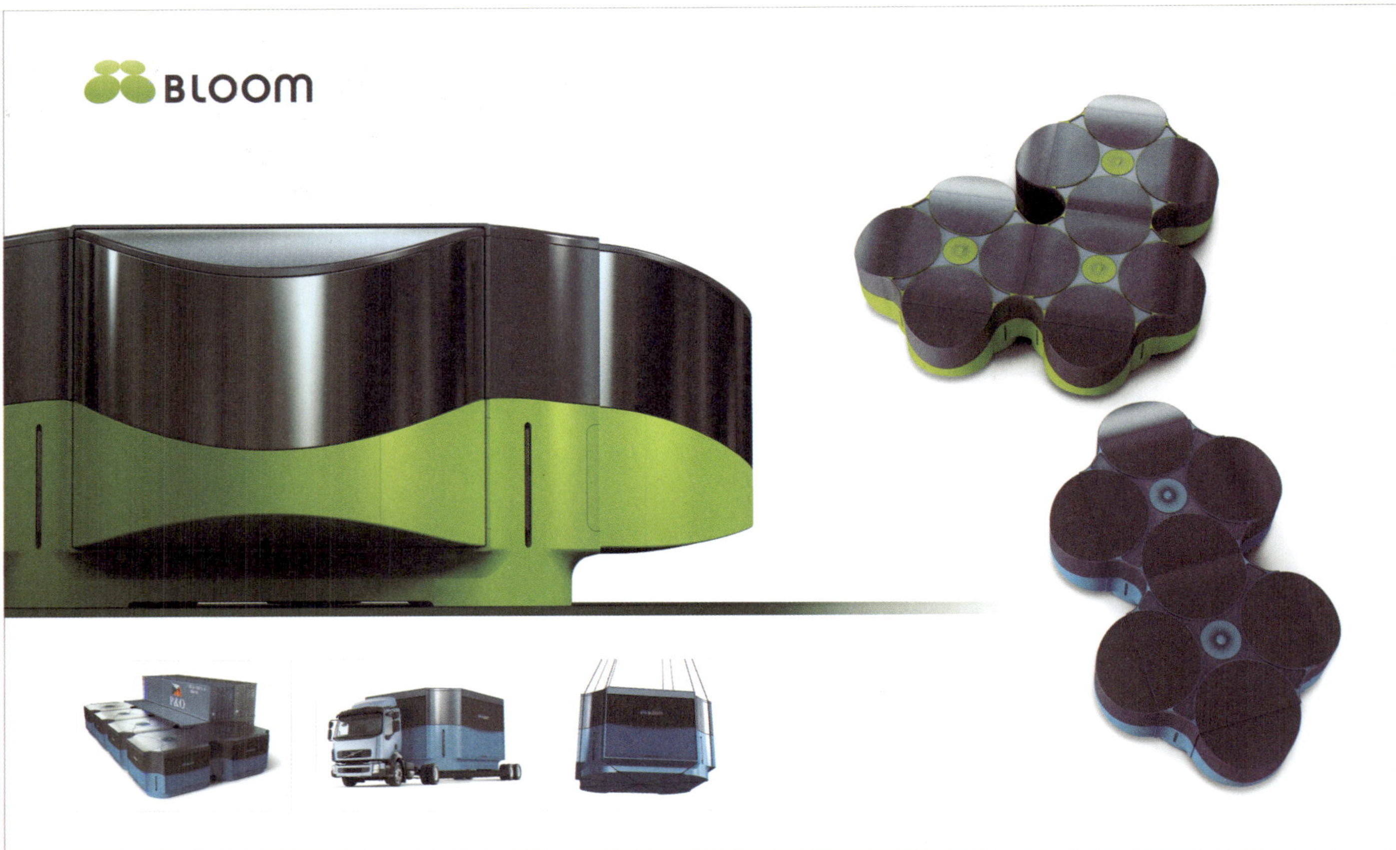

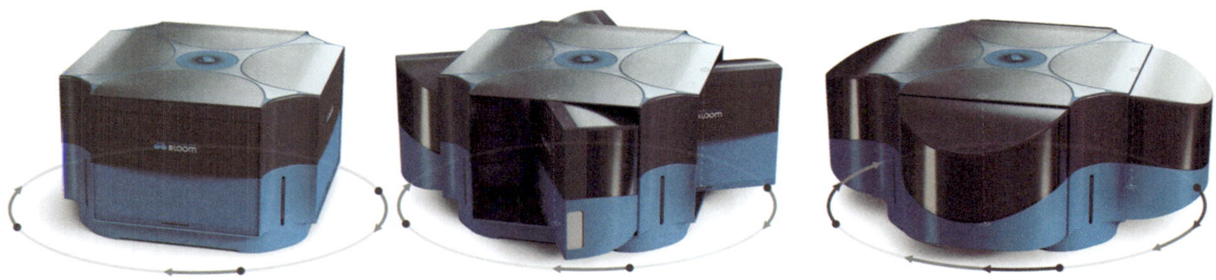

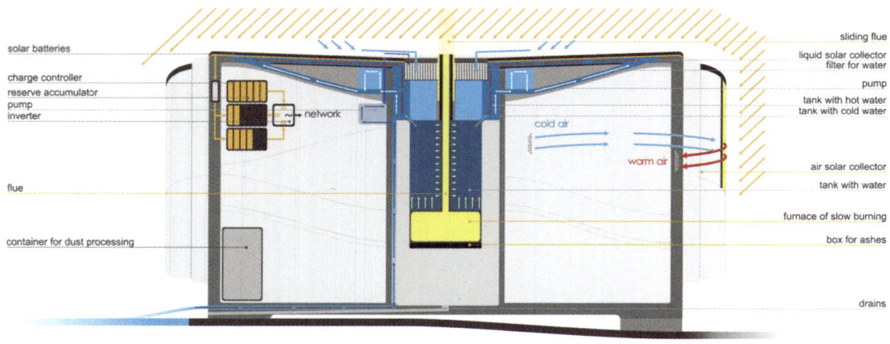

solar batteries
charge controller
reserve accumulator
pump
inverter
network
cold air
warm air
flue
container for dust processing
sliding flue
liquid solar collector
filter for water
pump
tank with hot water
tank with cold water
air solar collector
tank with water
furnace of slow burning
box for ashes
drains

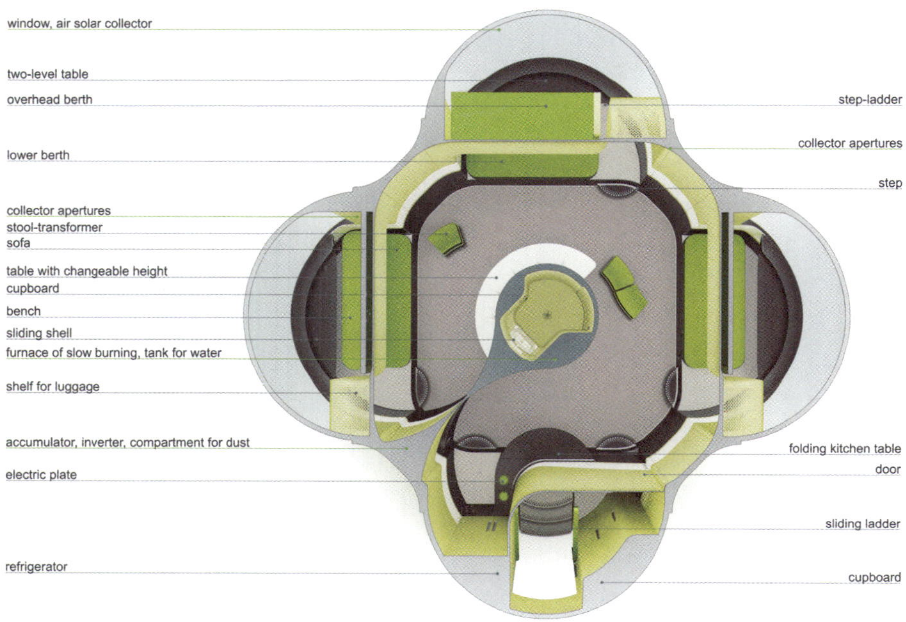

window, air solar collector
two-level table
overhead berth
lower berth
collector apertures
stool-transformer
sofa
table with changeable height
cupboard
bench
sliding shell
furnace of slow burning, tank for water
shelf for luggage
accumulator, inverter, compartment for dust
electric plate
refrigerator
step-ladder
collector apertures
step
folding kitchen table
door
sliding ladder
cupboard

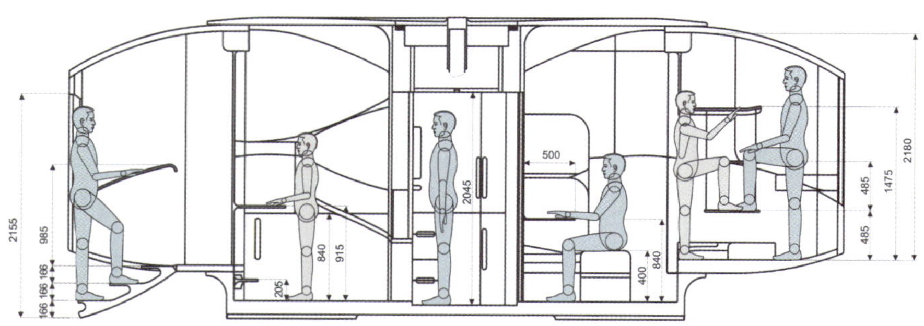

2155
985
500
2045
840
915
205
400
840
485
485
1475
2180

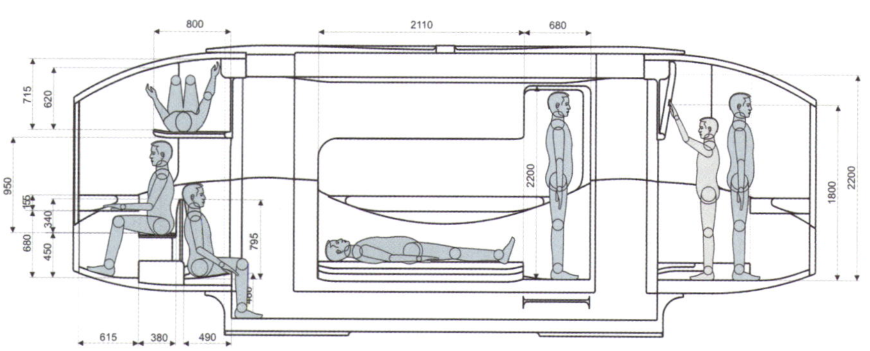

800
2110
680
715
620
950
340
680
450
795
2200
1800
2200
615
380
490

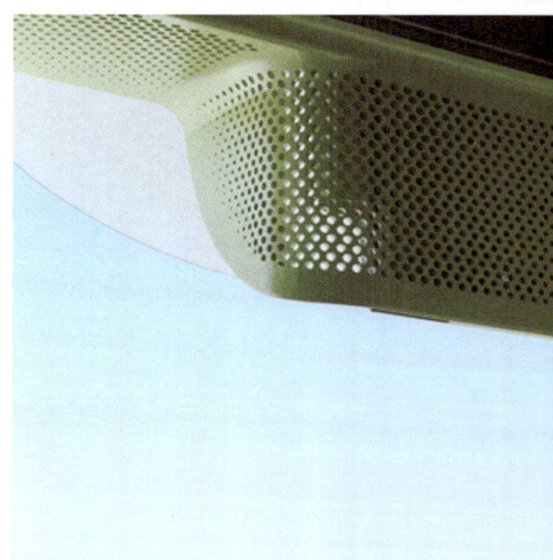

BLOOM

Crimean Pinecone Lamp

Designer Pavel Eekra
Photographer Alexander Nazaretski, Yana Tulaeva

The Crimean Pinecone Lamp consists of 56 plates and screws held without an internal frame. Light passes through gaps between the rounded forms of the plaques, which are slightly transparent themselves. This creates a special ambient pattern of light from the outside, while still exuding a practical downward-directed brightness. The lamp, which comes in a flat pack, has a natural wood veneer (maple), which further emphasizes the Crimean Pinecone's fusion of nature and technology.

Flataalto

Designer Ronen Kadushin
Client Open Design-Ronen Kadushin
Photographer Chanan Strauss

Flataalto is the abstract and modern cover version of the Savoy-vase. This world famous glass vase with the contours of a Finnish lake was designed in 1936 by Finnish architect and industrial designer Alvar Aalto.

The Flataalto flower holder has the characteristic contours of the Savoy-vase and in this version, made from a single piece of laser cut stainless steel, which is folded open before use. With a normal drinking glass in between, it is a minimalistic design vase with a wink to the design classic of the Finnish master. Arranging flowers is simple thanks to the rim's flowing form.

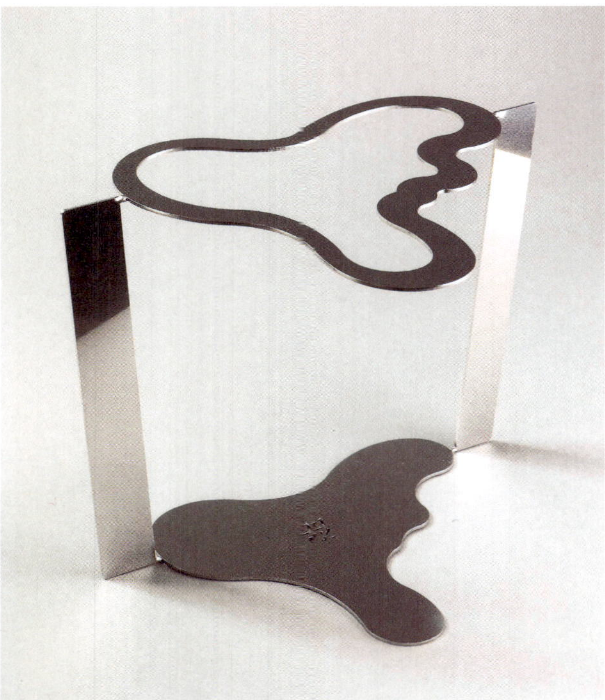

Vague Chair

Designer Ronen Kadushin
Client Open Design-Ronen Kadushin and Appel Design Gallery, Berlin
Photographer Chanan Strauss

The Vague Chair, with its unresolved and restless design, is part of "Recent Uploads" collection of chairs, an upbeat design exploration into the opportunities Open Design presents. The playful mode of the development process invites broad expression, knowing that there will be an effortless transformation of almost any design into a real chair. The incentive to experiment is pushed even further as these Open Design chairs are shared as a source for others to incite their own creativity.

Each chair draws its inspiration from different narratives: design references, emotional states, city life, and street art. But as a group they realize a personal design approach that embraces free improvisation and an immediacy of out comes, whether a computer image or a physical object. This mood resonates in the choice of raw cut, exposed material chairs.

The chairs are laser cut from a 6mm aluminum sheet, bent and assembled by hand. Bending a piece this thick is made easy using a hallmark detail for mulated a few years back; the integral twist hinge, which makes it possible to plan an exact bending position in a computer drawing, and retain the structural integrity of the chair, as part of the cutout. The selection of chairs consists of three unique structural configurations, each presented in two variations.

The designs of the chairs were recently uploaded onto my website for any one to copy, produce or experiment with. But you are also welcome to buy the chairs, produced and signed by the designer, at Appel Design Gallery, Berlin.

Blumen

Playful

Designer Måns Salomonsen
Client School project
Photographer Måns Salomonsen

Playful is a stool with a moving seating surface which creates a more activated sitting posture.

No screws or glue are used in its construction. It is all held together by the elastic band and the weight of the person using it. The band makes the seat surface bounce back to a horizontal position when it is not in use. The elastics also create a see – through surface that gives the stool volume and the option to create different looks.

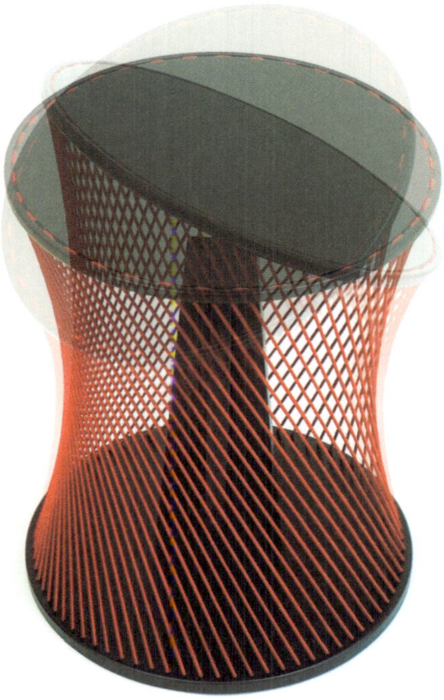

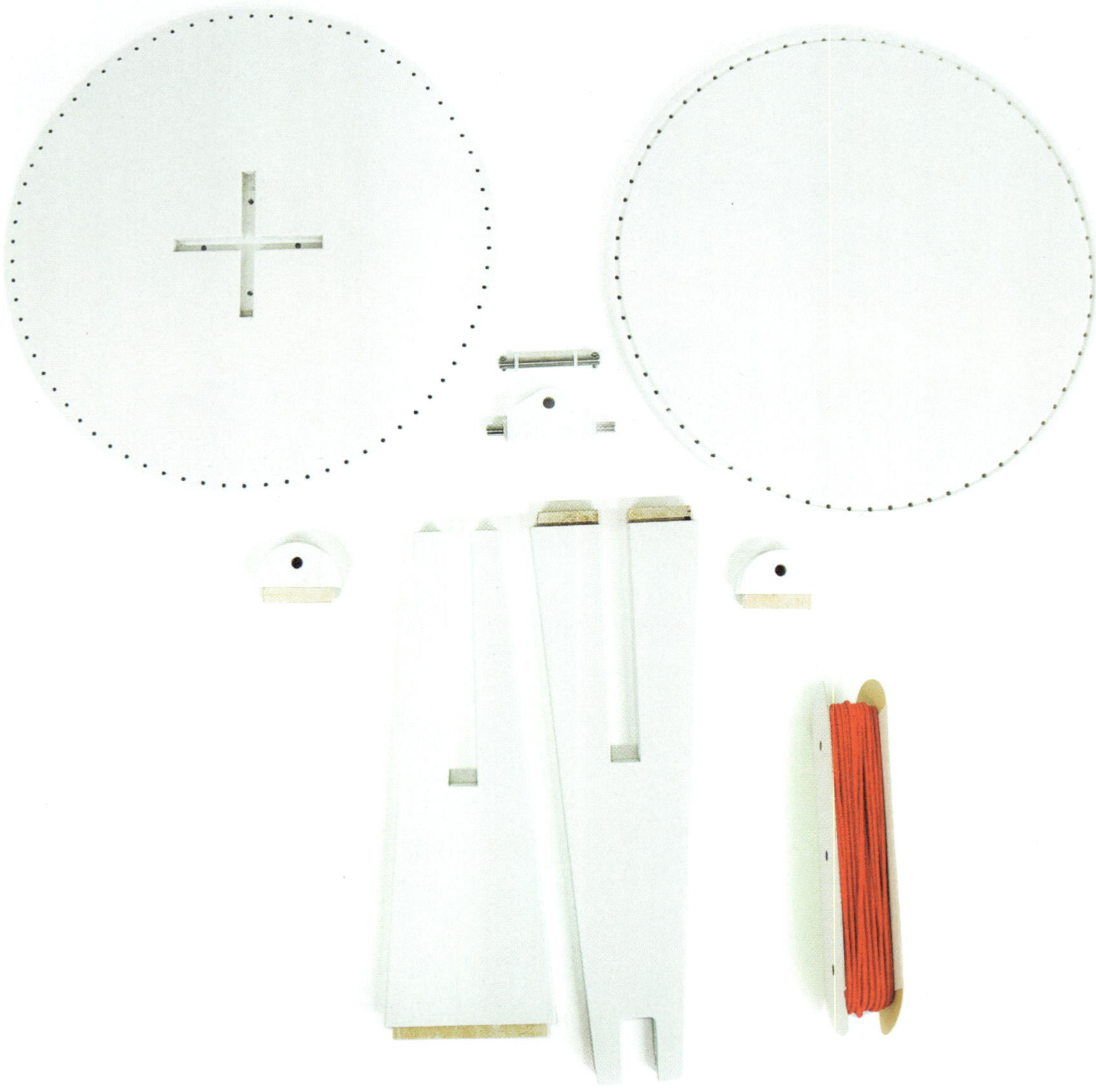

News

Design Agency Studio Veríssimo
Art Director Cláudio Cardoso
Designer Cláudio Cardoso and Telma Veríssimo
Client Studio Veríssimo
Photographer Ricardo Faria

News is a basket made of recycled 60 sheets of newspaper.

This resistant container is made in a traditional basket weaving technique, communicating very directly possibilities of domestic recycling.

Spoon

Design Agency	Studio Veríssimo
Art Director	Cláudio Cardoso
Designer	Cláudio Cardoso and Telma Veríssimo
Client	Studio Veríssimo
Photographer	Ricardo Faria

This is a lamp made of plastic spoons used in the coffee which are going to the garbage daily by the thousands because they are only used once. The objective was with this waste to create a luxurious lamp.

Spoon Lamp

Design Agency Studio Veríssimo
Art Director Cláudio Cardoso
Designer Cláudio Cardoso and Telma Veríssimo
Client Studio Veríssimo
Photographer Ricardo Faria

This is a lamp made from plastic spoons used in the coffee which are going to the garbage daily by the thousands because they are only used once. The objective was with this waste to create a product line, spoon collection.

Spoon Table

Design Agency Studio Veríssimo
Art Director Cláudio Cardoso
Designer Cláudio Cardoso and Telma Veríssimo
Client Studio Veríssimo
Photographer Ricardo Faria

This is a table made from plastic spoons used in the coffee which are going to the garbage daily by the thousands because they are only used once. The objective was with this waste to create a product line, spoon collection.

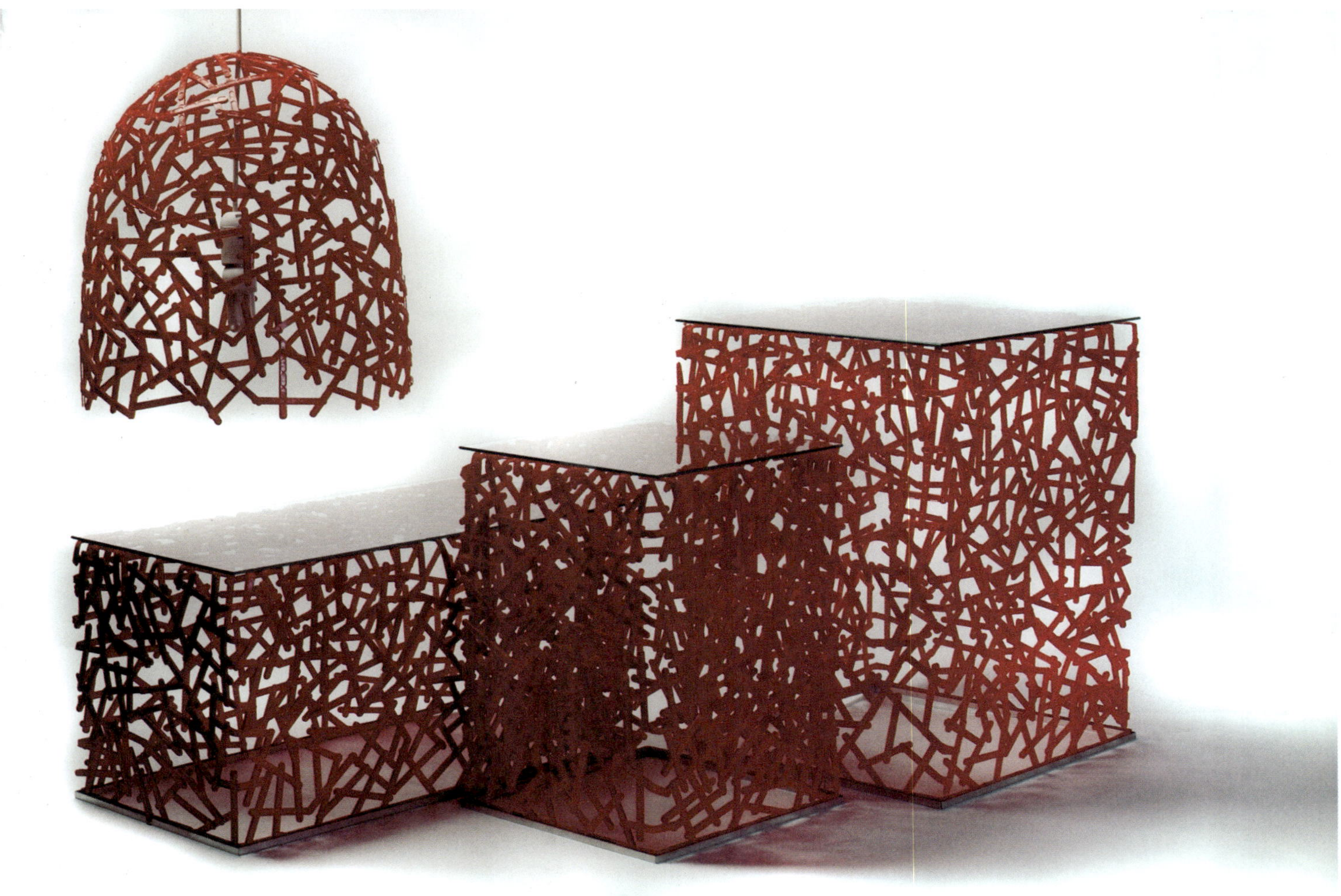

Vertical

Design Agency	Twentyone Studio
Art Director	Paolo Ciacci – Filippo Pernisco
Designer	Paolo Ciacci – Filippo Pernisco
Client	Fabrioso
Photographer	Manuela Tognoli

Last innovation in the garden architecture, which unfolds down spiral-shaped vertical lines creating a strongly expressive visual game of green falls. This is Vertical. Vase in fiberglass realized in a range of shiny powerful colours which capture light and shadows, perfectly matching and dominating every environment.

21
Vertical
Filippo Pernisco_Paolo Ciacci
Italy

Sunrise

Design Agency	Twentyone Studio
Art Director	Paolo Ciacci
Designer	Paolo Ciacci – Filippo Pernisco
Client	Simeg Marmi
Photographer	Livia Patta

Seats designed for urban design, parks or private gardens. And essential reading lines that adapt to any green space. Sunrise looks like a marble slab suspended from the ground with cuts to form blood vessels that allow plants and trees to surround the seat, creating an oasis in the countryside.

Green Gate

Design Agency	Twentyone Studio
Art Director	Paolo Ciacci
Designer	Paolo Ciacci – Filippo Pernisco
Client	Safa Arredamenti
Photographer	Livia Patta

Green Gate is an object/system composed of simple "frame" of steel that draws the common space by integrating with the local architecture and urban planning

The simple modular system allows to structure spaces, both public and private, in a completely innovative and multifaceted way.

Graph Bag

Design Agency Yuruliku
Designer Koushi Ikegami, Kinue Oneda

This is a pie graph bag.

Whatever surrounds us will turn out "round" right in a pie.

65

65

Green Marker

Design Agency Yuruliku
Designer Koushi Ikegami, Kinue Oneda

This is a grass-shaped page marker.

Marks you make appear and cover like new grass on your books and reference materials. Randomly placed "Green Marker" accumulates and spreads along your knowledge and curiosity.

Feel its growth as if letting nature take its course.

PEPPERS
271
was sent to me from Port Dickson
Malay Peninsula. This insect ate the
leaves and shoots, and though it did
the plants, prevented their bearing.
was 1 in. long and very thick and slug-shaped
all but the head, was bluish white, covered
tufts of blue or white spines. Its head
are dirty yellow, and at its tail above was
patch with four black velvety spots in a
The spines on its back are poisonous,
like all of this set can sting rather
I failed to rear any of the moths
as they were too much
have often seen them
is certainly a
stinging

Beetling

Design Agency Design nobis
Designer Hakan Gürsu, PhD

Beetling is an indoor seating system that has its own advertising and information area. By its elegant transparent body, an inner place is created for posters which would be in sight of any person passing by or intending to sit, for certain. Two transparent layers of the body holds the poster still, also avails the replacement of the poster when needed. Not only it is a perfect place for attention grabbing advertisements, it is also excellent for information and signage use.

Beetling can be used in any waiting area such as banks, hospitals, airports, by simply arranging the seating units along with the intermediary service tables. With its fine looking profile, it is fit for all modern architectural environments.

Beetling brings its users many ergonomic features. The rubber disks of seating surface serves to skin refreshment and increase comfort level. The perforated surfaces created naturally by these disks help ventilation. Also it prevents a slippery surface and massages the sitting person.

Design is awarded from Bureau of European Design Associations with an A'Design Award in 2011.

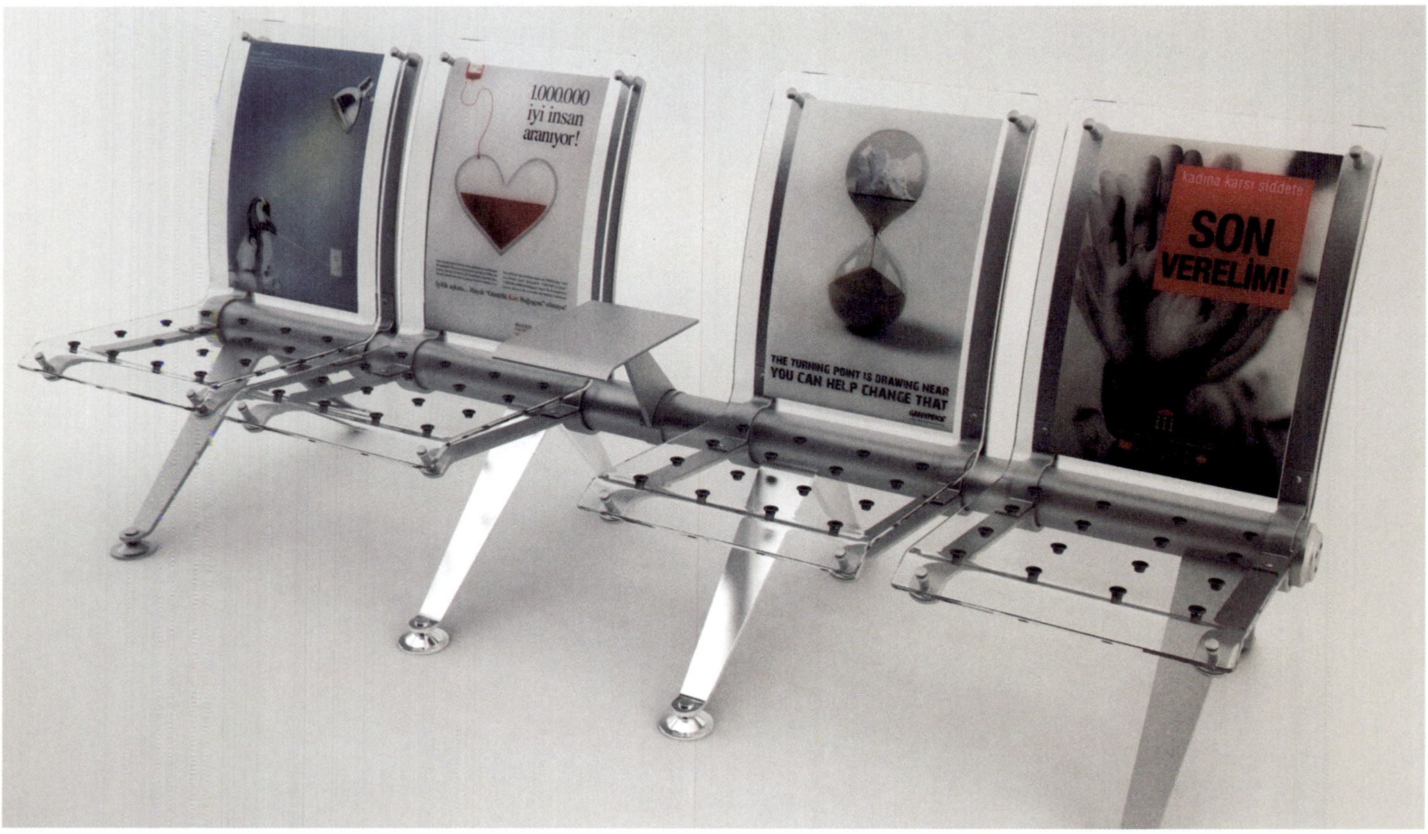

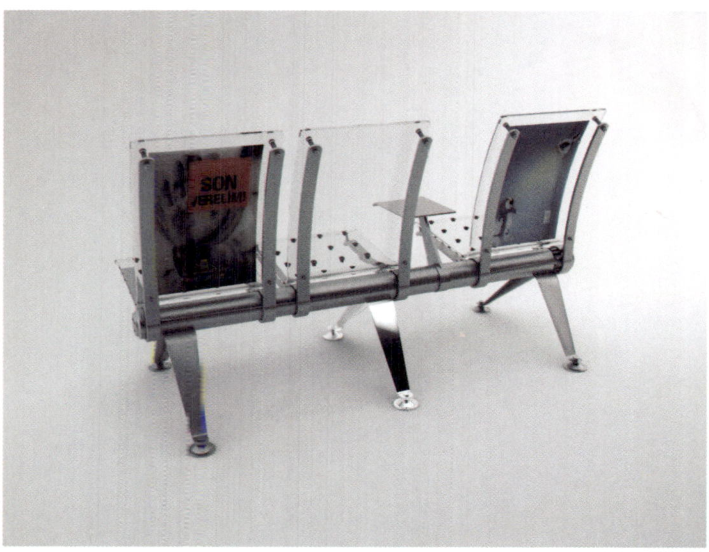

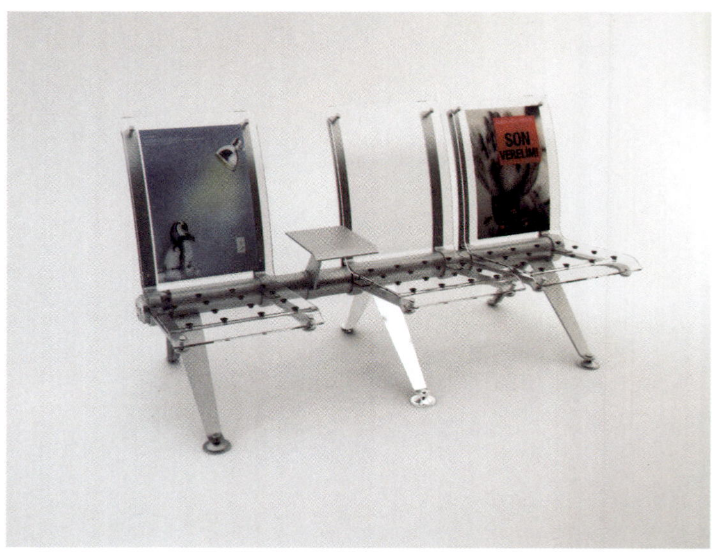

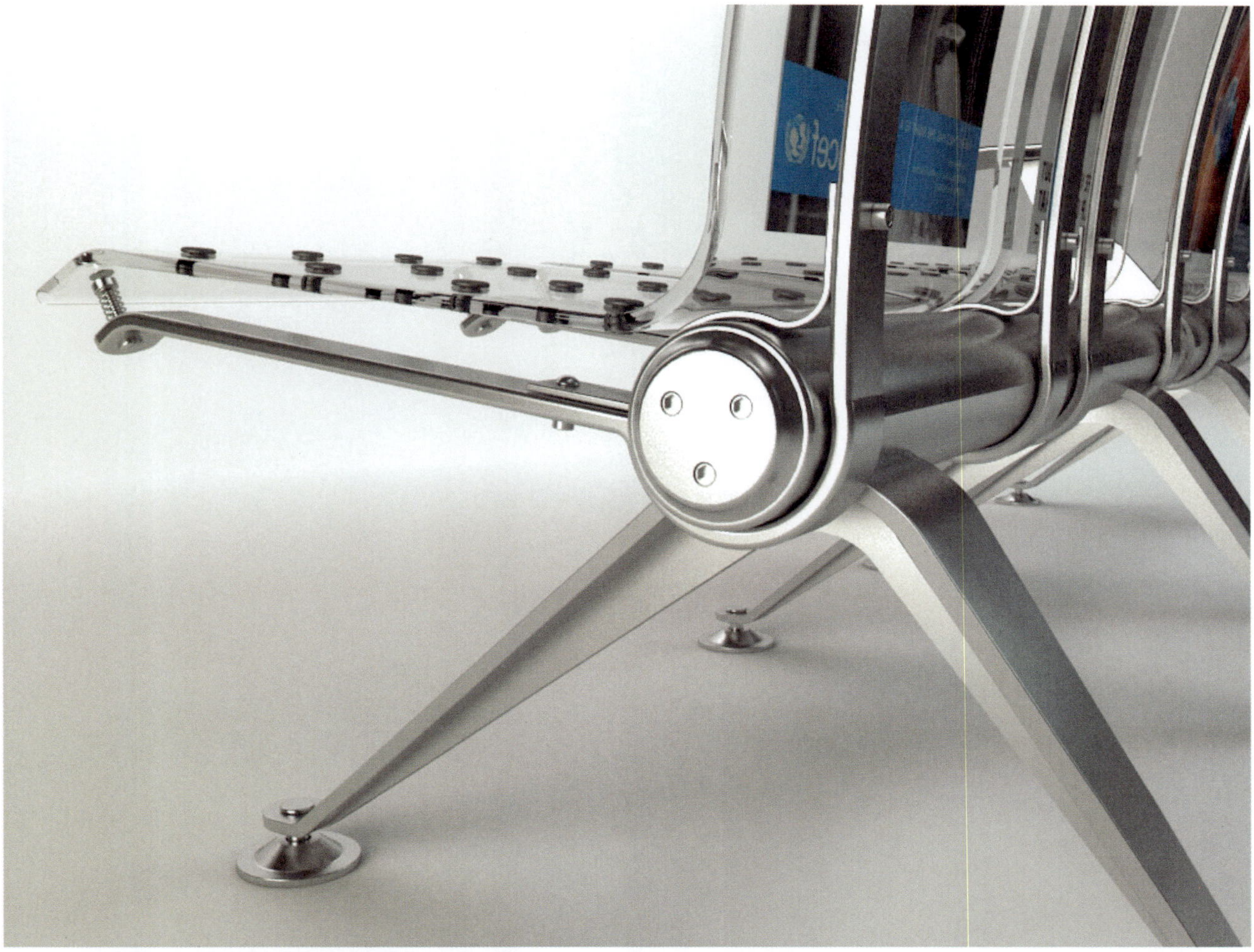

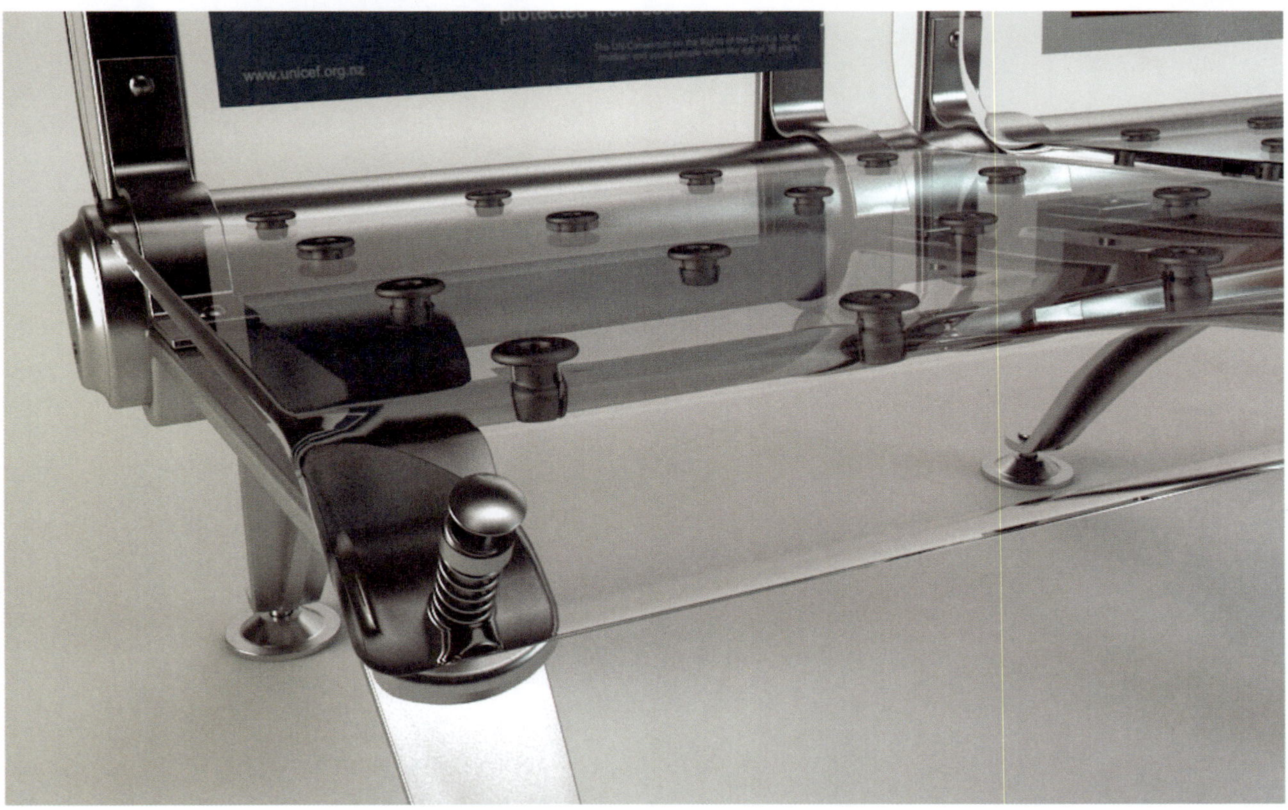
www.unicef.org.nz

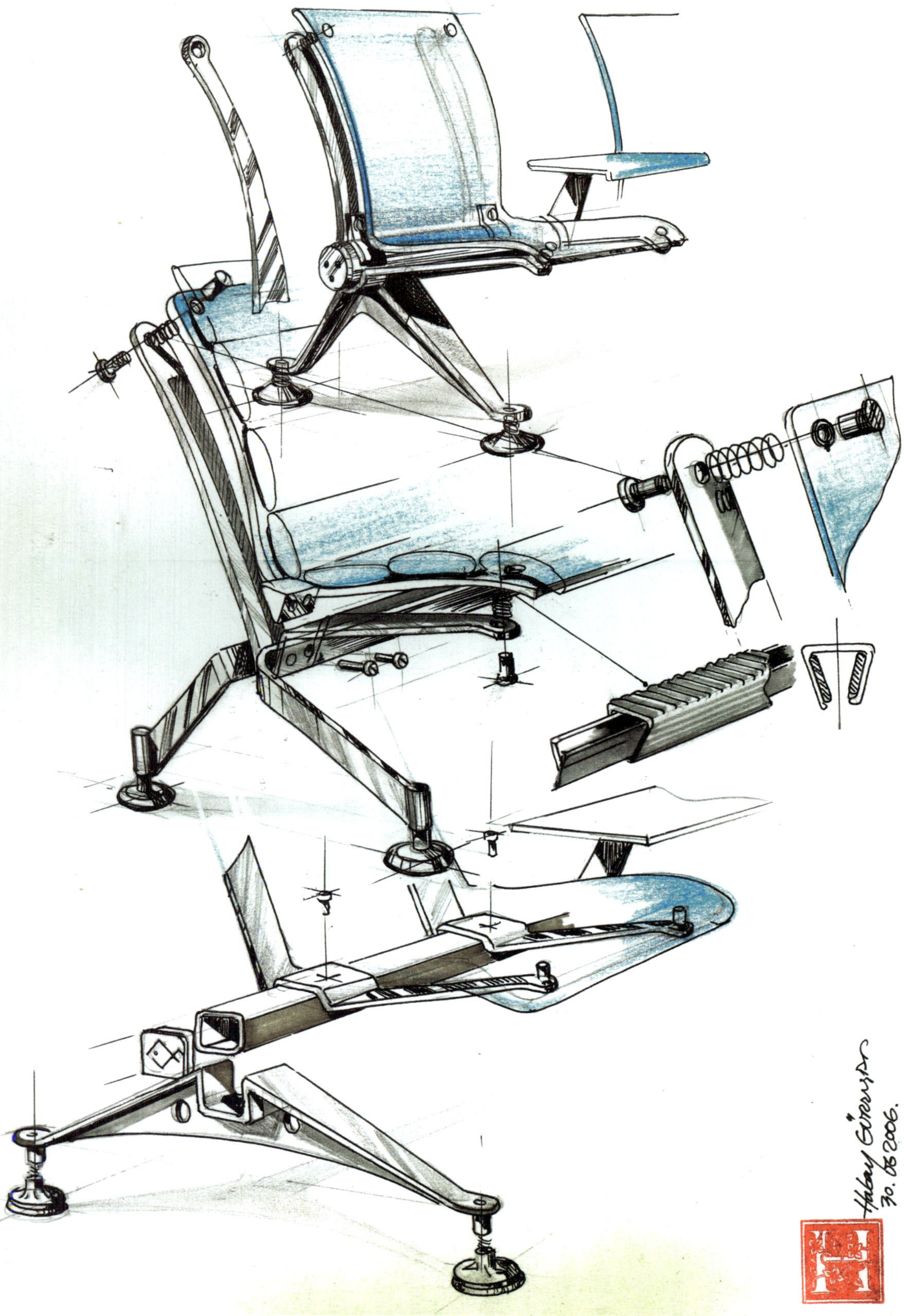

OUR WAY OF LIFE IS RUNNING OUT
YOU CAN HELP CHANGE THAT
unicef

2D LED

Designer DING3000
Producer Skitsch

The quick hand sketch of an archetypical lamp was the inspiration for "2D LED". By means of modern LED technology it is possible to keep the body of the lamp as thin as possible, which gives the luminaire the appearance of a sketched line. The sheathed metal hose within the foot provides an adjustable hood. "2D LED" is available as table or floor lamp.

In 2011 the 2D LED family was enlarged by a pendant version (presented at the Salone Del Mobile 2011).

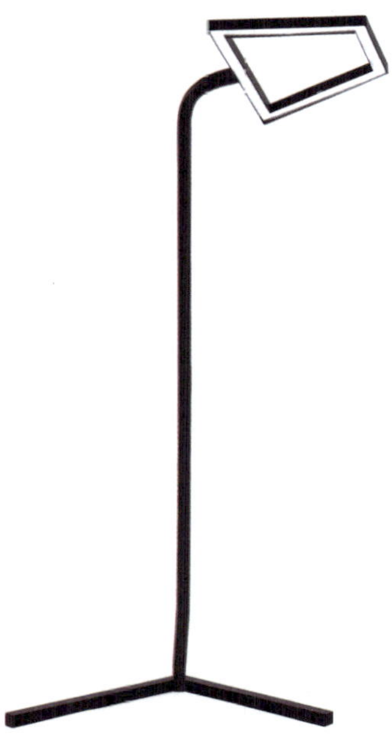

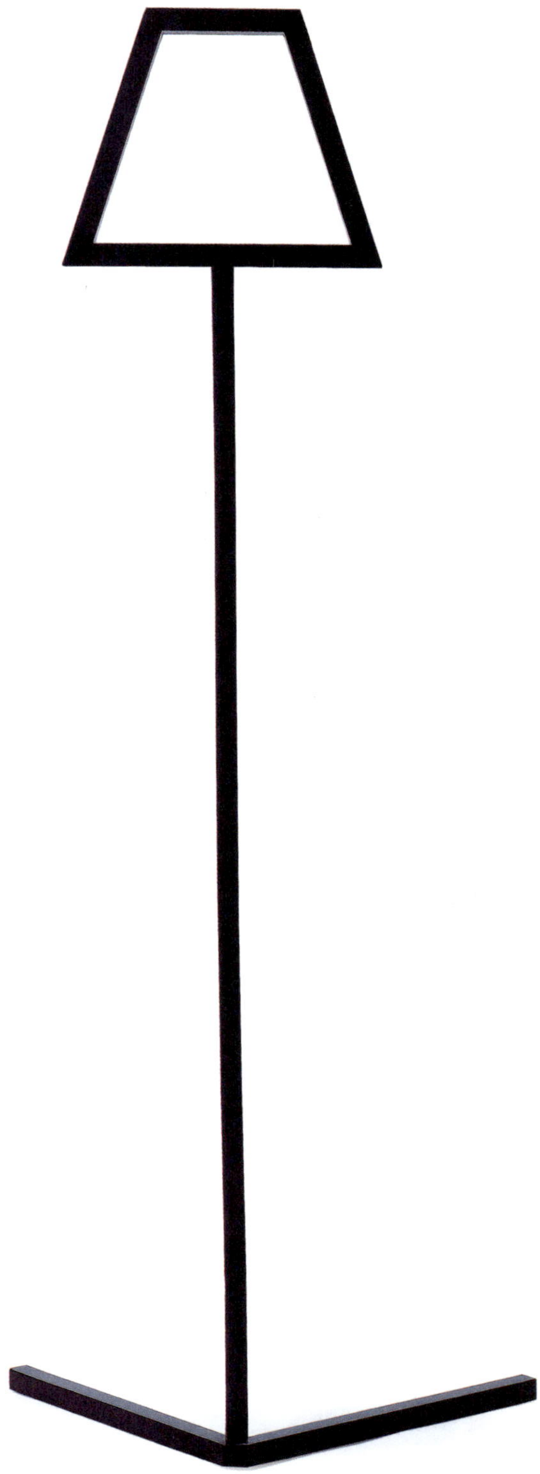

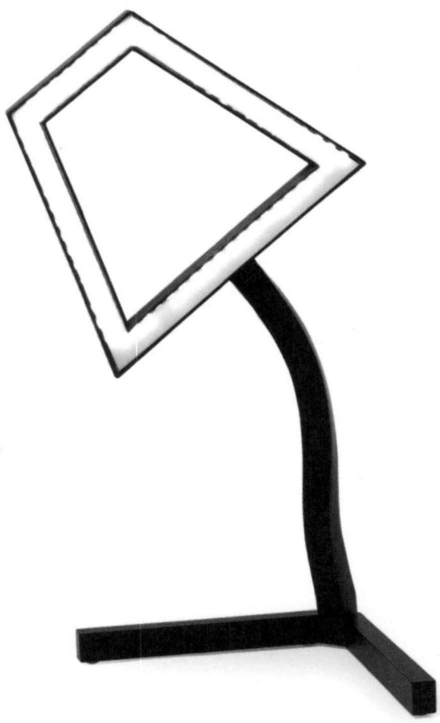

SL17 BASKETBIN

Designer DING3000
Country Germany

Garbage separation has never been sexy–up to now. BASKETBIN takes care of a clean separation of wet and dry materials without visually polluting the environment. This bin within a bin is suitable for every office, and it makes quite an impression altogether.

Wet and sticky garbage disappears discreetly under the lid of the small dishwasher proof bin and can easily be carried to the nearest dump. Behind the fence of the big bin papers and plastic wastes are separate.

reddot design award
winner 2010

product
design
award
2010

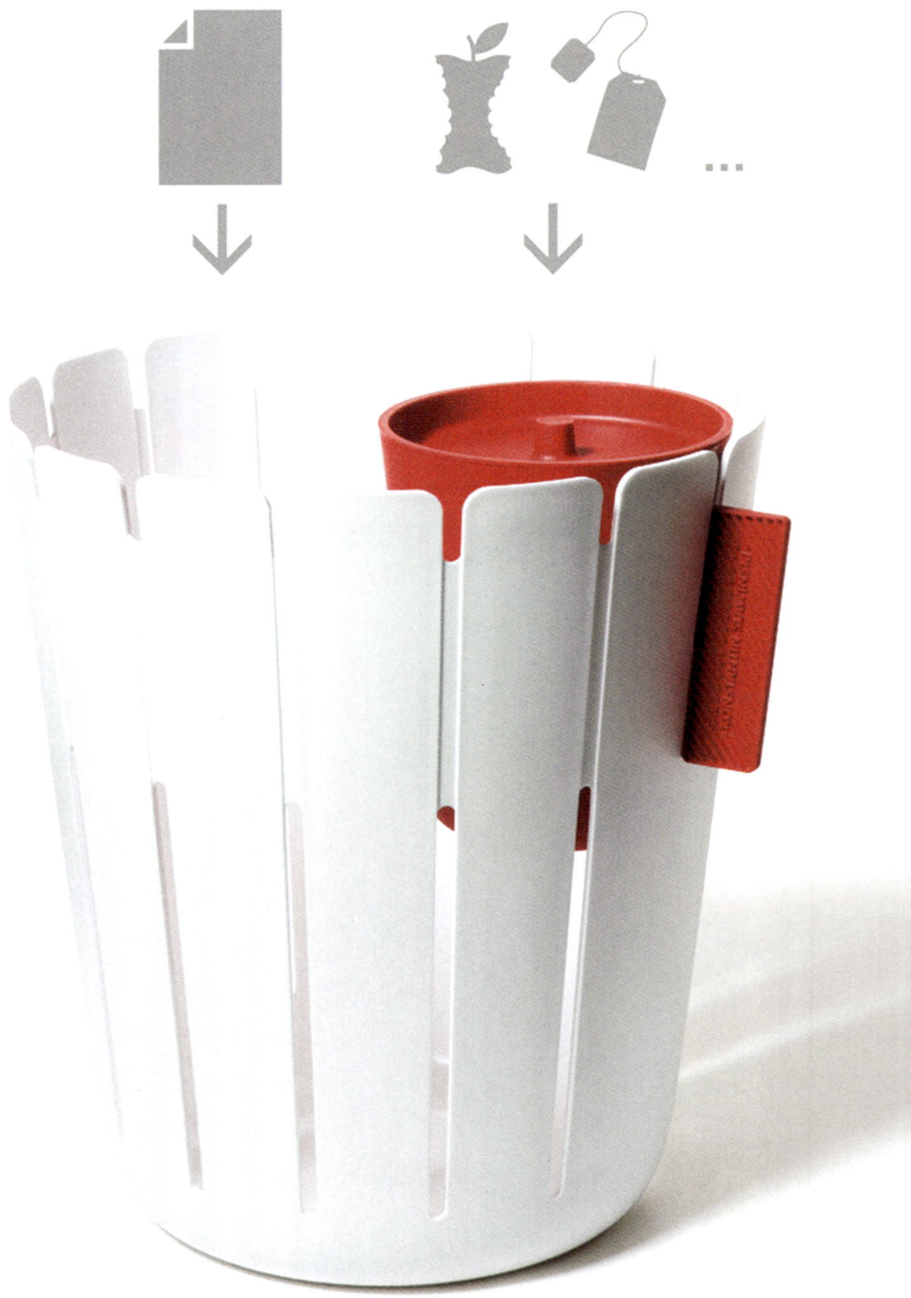

Corona

Designer DING3000
Producer AUTHENTICS

Although the colour of this wall clock´s face and hands is plain pure white, it is still an eyecatcher. An intense colour on the reverse of the clock´s hands and face makes appear a colourfully glowing corona and enables you to clearly ard well read the time of day.

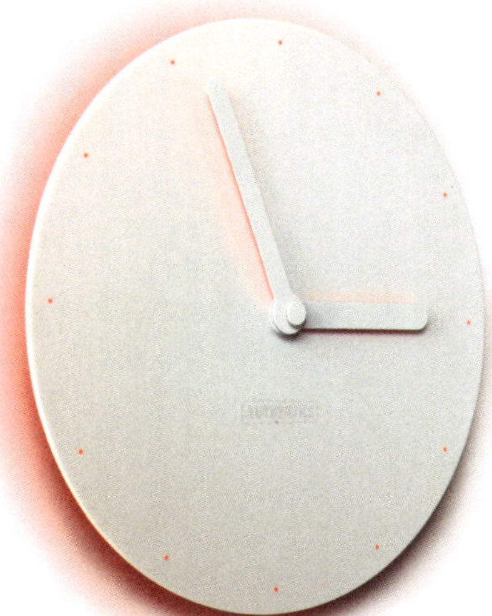

DING3000's Animal Tales!

Designer DING3000
Photographer Philipp Nemenz

During the Milano Design Week 2008 the German design studio DING3000 (Carsten Schelling, Sven Rudolph, Ralf Webermann) presented its new collection themed Animal Tales. It's a set of furniture, luminaries and accessories all inspired by the animal kingdom. The whole collection received a special mention of the Design Report Award which takes place at the Salone Satellite since nine years ago. The enthusiastic jury was formed by Yves Béhar (CH), Frédéric Dedelley (CH), Franke Elshout (NL), Roberto Feo (E), Harri Koskinen (FIN) and Lars Quadejacob, editor-in-chief of the design report.

Elegant afghan hound–floats swiftly through every living room! The sofa tazi is inspired by an afghan hound. A curtain of white fringes covers substructure and legs. Only the seat remains visible. Due to this design-trick the sofa seems to "float" above the floor.

INANDOUT

Designer DING3000
Producer Konstantin Slawinski - Housewarming Objects

It's up to you to decide which form the trivet is to take: in or out, round or jagged, suitable for small or large pots. The trivet INANDOUT suits your request in a vers flexible way. No wonder since it consists of soft silicone and can be turned with one flick of your wrist. Concerning the colour, your wish is also our command. INANDOUT comes in blackberry, green, blue or grey colours. With this concept of freedom of choice one thing remains 100 percent stable: your pot remains on the table.

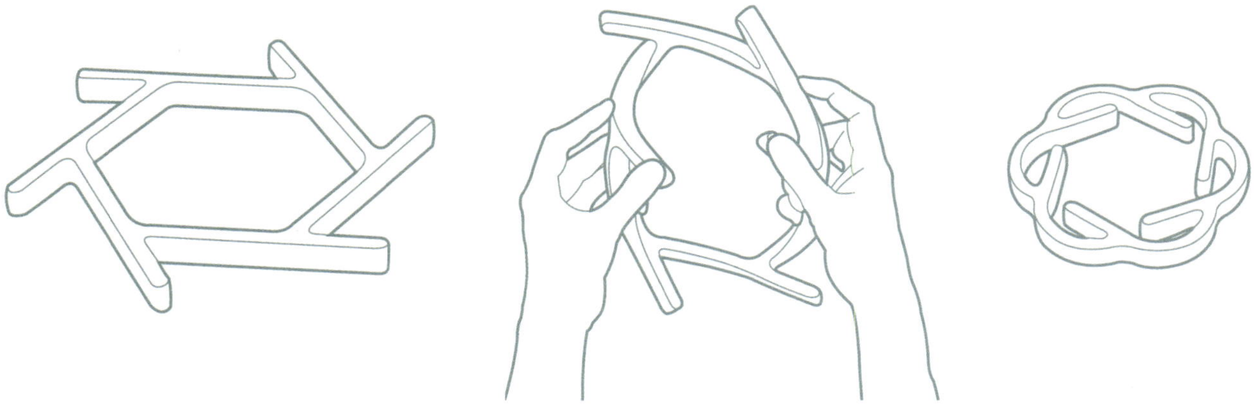

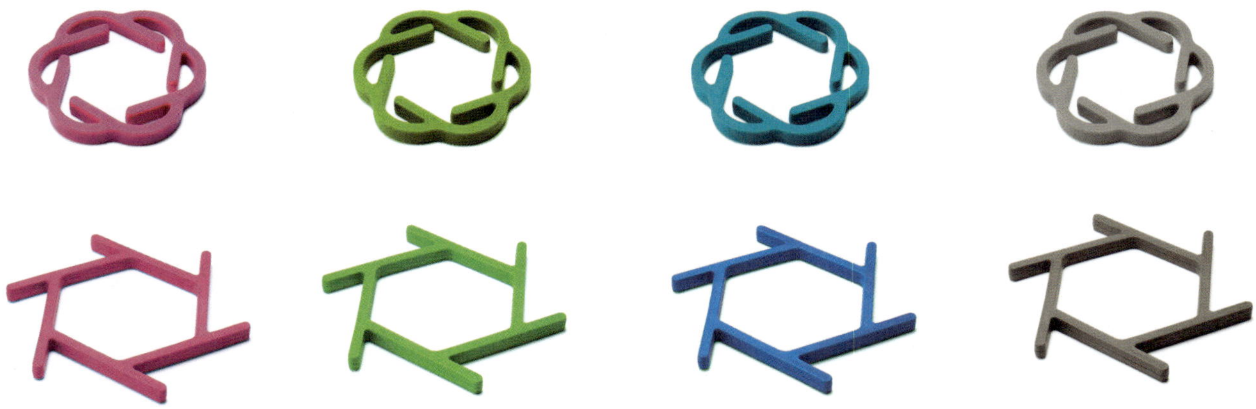

JOIN

Designer DING3000
Producer Konstantin Slawinski | Housewarming Objects by Slawinski & Co. GmbH

Cutlery turns into cover. JOIN is more than a simply knife, fork or spoon. It is a decoration for the table. The magic joining mechanism fascinates everybody. But not everybody will manage at once to transform the little sculpture into cutlery. A little skill and a good eye is necessary.

But do not worry, up to now nobody starved when unravelling the magic knot. Many times a meal grew cold because the cutlery itself, made of long life high-tech plastic material, was simply too fascinating. You should book JOIN a place at your table.

The cutlery was developed in co-operation with BASF.

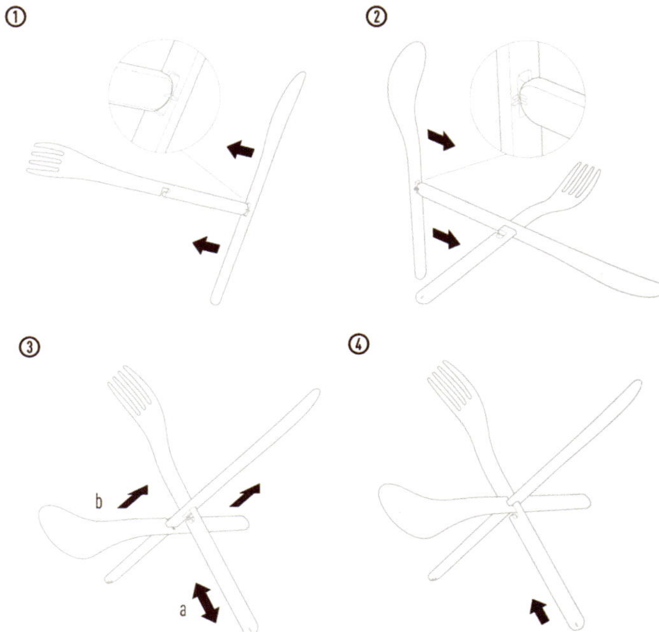
①
②
③
④
a
b

JOIN TABLE

Designer DING3000

The inspiration for the table-top support of JOIN TABLE was an ancient Japanese game. Three legs can be fitted into each other without any tools by means of the simple but enigmatic principle of the devils knot and merge to one mysterious object.

JOIN TABLE is the big brother to our cutlery JOIN, which is based on the same principle.

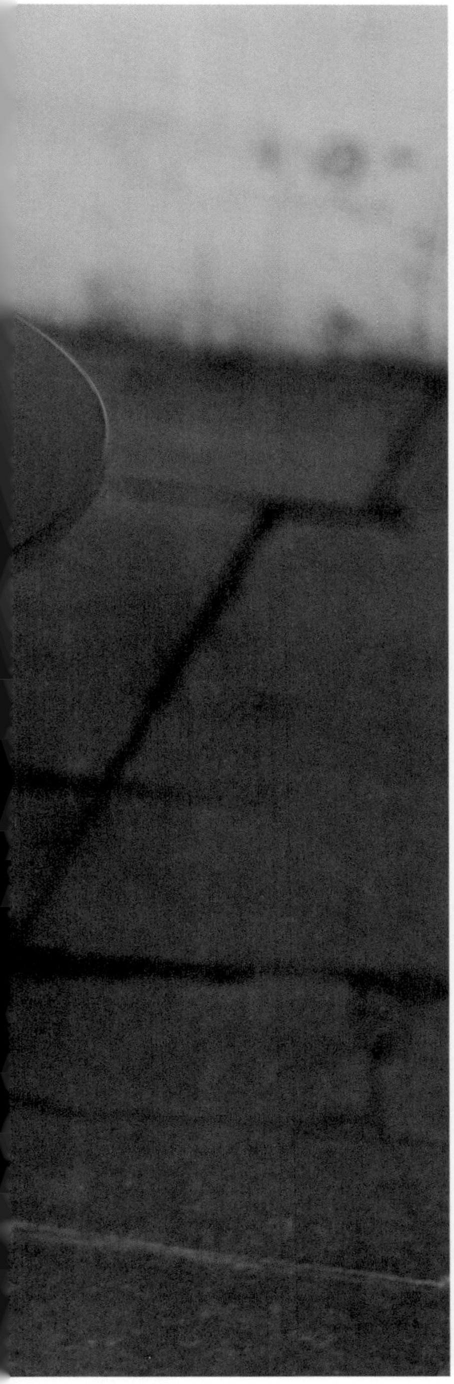

Knickding

Designer DING3000
Producer PULPO

The angular edging changes a simple sheet of metal into an exciting shelf.

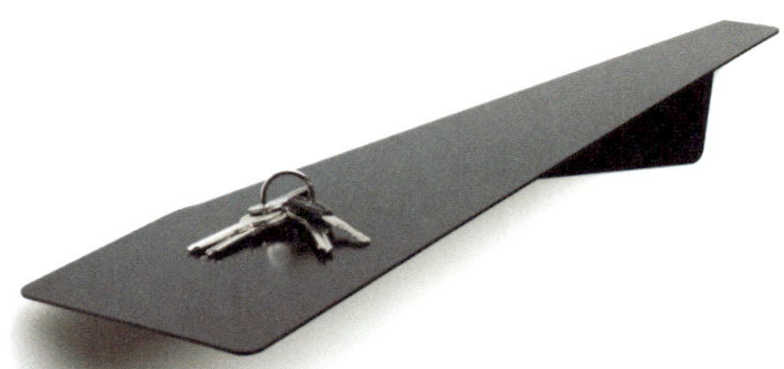

LUCKEY

Designer	DING3000
Producer	FRED & friends

Bottle-opener with the shape of a typical key. A give-away with the right turn.

Kerzenhaltding

Designer	DING3000
Producer	Sompex

Kerzenhaltding is a candleholder, which simply uses the holding force of a power socket.

The correlation between electricity and light is supposed to be irritated by that concept. The lit candle paradoxically emits light, although the plug does not transmit any electricity.

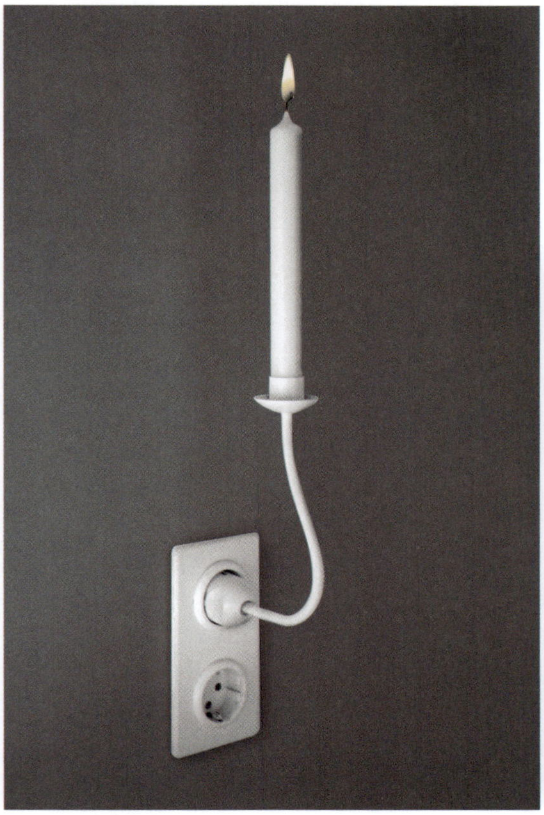

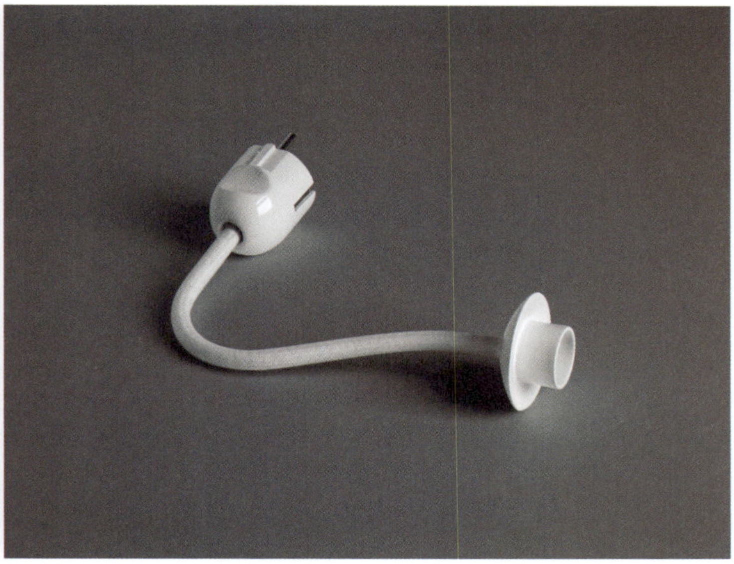

Multistar - Candleholder

Designer DING3000
Producer Sompex

As a single or in a formation, multistar always cuts a fine figure! The candleholder can be arranged in different geometrical patterns to fit all sizes and shapes of tables.

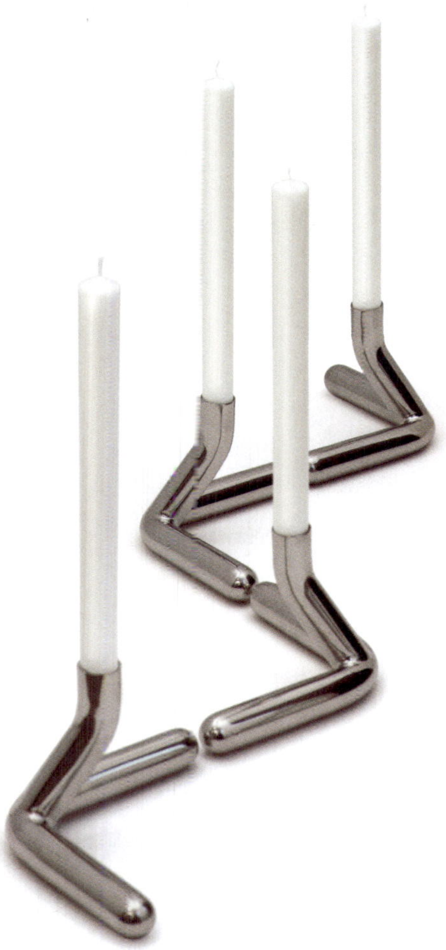

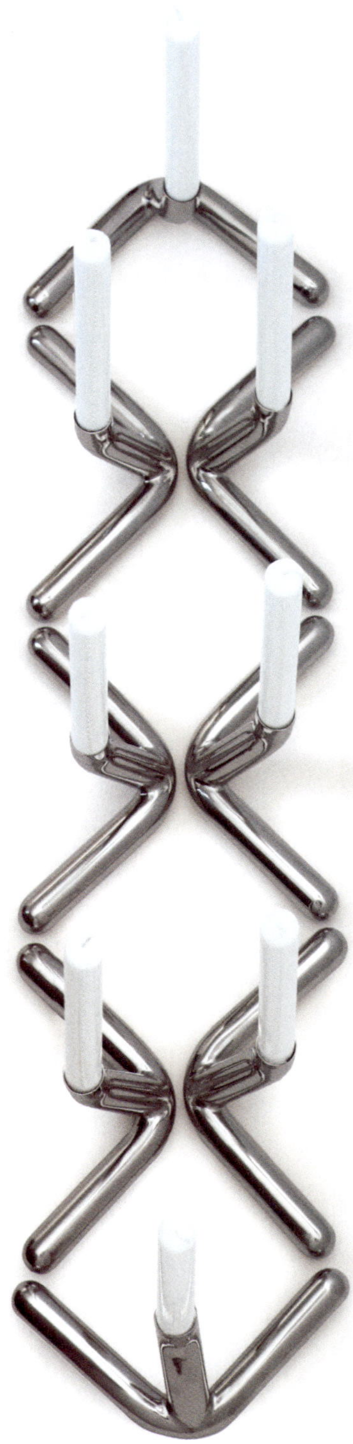

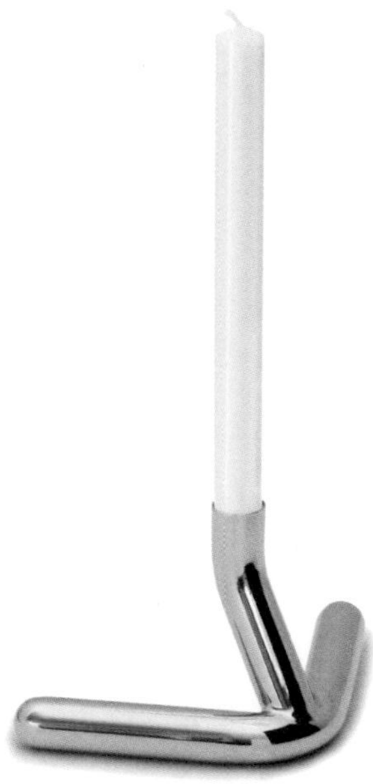

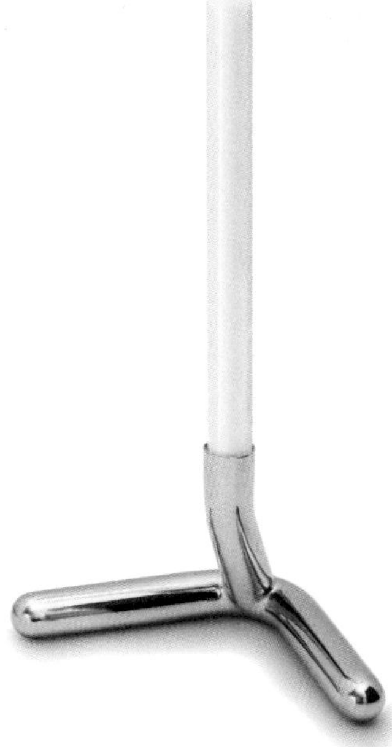

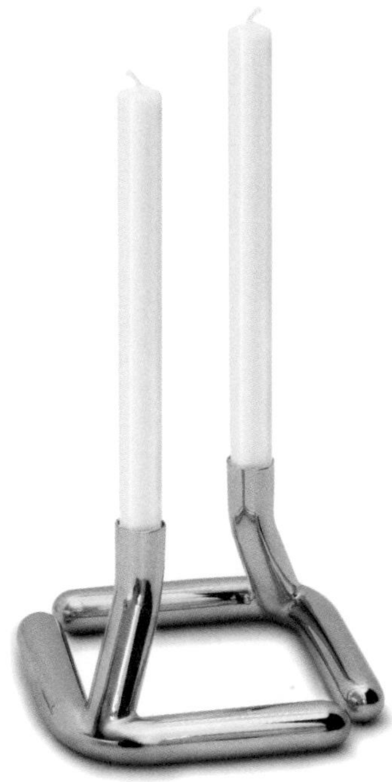

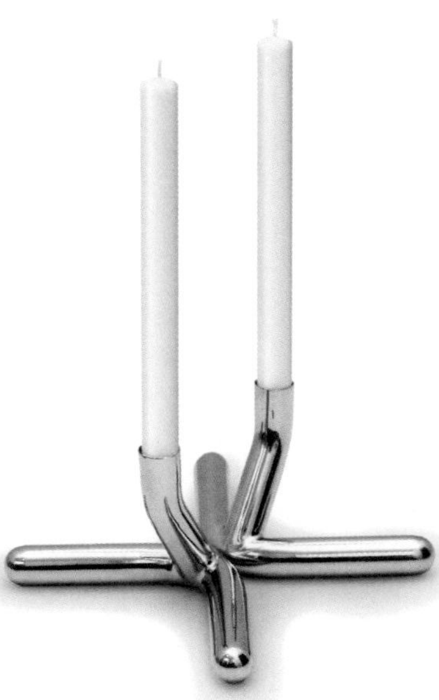

Nooodle

Designer DING3000
Producer Konstantin Slawinski - Housewarming Objects

Noooodle combines two functions with a playfull approach. It is a spaghetti measure for 1 to 4 portions and at the same time a casserole stand. Inspired by the image of a single spaghetti, the loops form the measure for four different sizes of portions.

BILLY HEIDENREICH

Designer DING3000

BILLY HEIDENREICH is a rack with a lectern. The book shelf is enhanced by exchanging the middle rack with an apparent and convenient add-on element. The lectern is modeled on classic examples. With its shelf-related design and workmanship, however, it is easily integrated into the overall picture.

OderSo – shelf-system

Designer DING3000

OderSo is a shelf-system consisting of a single module. The geometry of this base-module in combination with the making of the edges renders the possibility of a multidimensionally modifiable and thus extremely variable assembly: by simplely piling equal modules can be assembled to sideboards, shelves or room dividers. According to the way of assembly the furniture's impression varies from orderly (in the sense of symmetrical/straight) to unorderly (in the sense of unsymmetrical/rampantly).

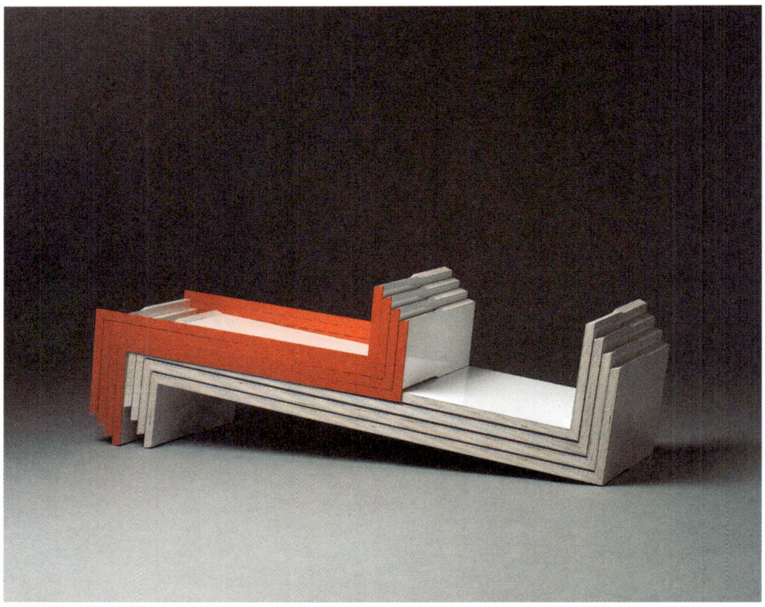

1000 chairs
TASCHEN
DESIGN
Deutschland
DESIGN
Skandinavien
Collapsibles

Block Shelf

Designer Green Furniture Sweden

Block Shelf can be hung on the wall or from the ceiling, where it also can have the function of a thin room divider. It comes configured for 4 shelves of 2 plank's width, but can also be mounted for 8 shelves of 1 plank's width.

T-shirt Chair

Designer Green Furniture Sweden

T-shirt Chair is like a large colourful rag rug, an easy chair made of left over clothes or rags. Every chair has its own unique colour and texture combination. The textiles can be individually replaced with worn clothes and textiles from your surroundings, and in that way the T-shirt Chair will evolve and collect history over time – an almost "eternal" piece of furniture... (The frame actually has life time warranty).

In the "Rag Edition" we upholster using left-over fabric pieces from the sofa manufacturing, in a selected "all-colour"-mix. Selected theme colour is also available, 998 Euro.

"Frame Only": T-shirt Chair is also sold as a plain frame, for you to fill rags and worn clothes, price 298 Euro.

"Special Edition": T-shirt Chair is also available in a numbered and signed "special edition" where the designer herself actually weaves using used T-shirts.

Leaf Lamp

Designer Green Furniture Sweden

BLOMSTERODLING

Agatha

Designer LZF

Agatha, created by Spanish designer Luis Eslava for LZF, is a hanging lamp whose shape is achieved through the concentrical superposition of materials. Same as in every LZF object, its lines remind us of those we find in nature: a flower, a tree, an animal. Light in aspect, Agatha suggests simplicity and originality. This lamp "blooms", forming a semi-spherical object, in two different sizes: 76 cm x 37 cm and 140 cm x 80 cm, or expanding to form a full sphere: 76 cm x 76 cm. Available in 8 different finish tones, Agatha can be adapted to a wide range of spaces: private homes, contract, ephemeral settings, and many more.

Air

Designer LZF

The Air collection was an idea of Irish designer Ray Power who, fascinated by the huge potential offered by polywood decided to explore its whole range of possibilities, manipulating and using the sheets to create tridimensional objects. The result, a conical stand and shade which suggest both visual and technical stability. LZF's intention is to achieve, with this design, a recognizable shape through the use of a material so wildly natural, yet, achieving a form of iconic expression of the most classical models. Its original shape is a good asset for every place: private homes, commercial areas, ephemeral settings, and so on. Air is the winner of the prestigious award Red Dot 2010 (Germany) and the Good Design Award (Chicago, 2009).

MGM

Armadillo

Designer LZF

Armadillo, is a design by Spanish designer Eslava. Its shape reminds us of that strange mammal from Southern America whose activity starts at sunset. Armadillo gives a warm light, almost as timid as this little animal's attitude at dawn. The lamp, woven in a coat of superposing scales, features a unique play of light and shadow. The textured material of the armour opens a whole universe of decorative ideas. Its sifted light gives the place a genuine warmness that proves perfect for public or private interiors.

Cactus

Designer LZF

Cactus is a likeable and elegant table lamp, created by Spanish designer Miguel Herranz for LZF. Its volume is one of the main features of the lamp, one that attracts attention even when it is switched off. When switched on, its beauty is boosted thanks to the interesting graphics achieved with the overlapping of the wood veneer. The bottom piece is flat and extremely thin. Cactus's merit is its ability to establish a dialogue between the soundness of its shape and the lightness created by it's small base. Dimensions for this unusual lamp are 17 cm x 37 cm.

Link Chain

Designer LZF

Following the shape and form of the original link, created in 2007 by Irish designer Ray Power and with the collaboration of Spanish designer Marivi Calvo the link chain is born! This member of the link family has grown, and considerably so, reaching lengths of up to 172cm. Carrying on from Ray's study of form and geometry present in its predecessors this stack of moebius strips comes in two standard sizes made up of 3 sets and 4 sets of the large link pendant modules. The result is a more powerful version, if possible of the large link suspension light, still using however, with great skill the polywood veneers with which it won the the design plus material vision prize back in 2007.

MiniMikado-A

Designer LZF

Its shade is made of several pieces of die-cut wood veneer which gives it a dynamic, easy and sensual appearance. The fringe creates a play of light and shadow, while its lightweight structure achieves an elegant, stylised shape. The MiniMikado-A version features a slightly modified design, making Mikado lighter and easier to adapt to the wall format. The result is a product with a great formal richness. The Mikado family was created by Spanish designer Miguel Herranz in 2005. It was awarded the Good Design prize in 2008, in the lighting category, and got the Design Plus award in 2006 ("Light and Building", Frankfurt). In the wall version, the MiniMikado dimensions are 65 cm (length) and 40 cm (width). Available also as floor version and hanging lamp.

INTERIORS
New York Interiors

Paisley

Designer LZF

Paisley was created by Spanish design studio Luis Eslava Studio for LZF Lamps. It is a modular screen whose geometric, curvy shapes transform the light that passes through it. Paisley is an architectonic piece that allows us to define spaces by assembling several modules, but also a canvas where intricate shadows are projected and turned into an ornament, changing shadows that adapt themselves to the direction of light. The inner part is manufactured in natural or dyed wood veneers, while the frames can be in cherry wood or white lacquered aluminium. It's dimensions are 2 m heigh and 90 cm wide.

Pod

Designer LZF

Designed by German Burkhard Dämmer, Pod is next in kin to hanging version Poppy, both of which represent the most natural version of this artist's achievements. Pod is more compact, elongated and its slats are less separated than in Poppy. Based on the arranged repetition of shapes a curvy plasticity is achieved whose sustained rhythm is strengthend by the light coming through the wood slats and projecting onto the desired surroundings. Pod is available in two sizes (47 x 49 cm and 53 x 60 cm). Its compact, oval shape reminds of a fruit or a seed.

STOP
GO BACK
KEEP
OUT

Saturnia

Designer LZF

Saturnia is a lamp family of clean, soft lines, made up of hanging lamps and designed by Basque-Swiss designer Óskar Cerezo for LZF Lamps. Its name and shape are reminiscent of Planet Saturn, the only planet whose rings can be seen from the Earth. On one side, the light passing thorough the wood sheets creates a warm ambience. On the other side, the plexiglas diffusor allows for a good working light. Moreover, it incorporates state-of-the-art electronic technology that assures a minimum electrical consumption and is a perfect choice for direct as well and indirect lighting, both in private homes or contract. Available in three sizes: as happens with other LZF lamps, it can be combined to create amazing lamp clusters. The dimensions are: 49 cm, 85 cm or 120 cm diameter.

Energie Beheer Nederland Meeting Room Utrecht Fokkema Architecten Photographer Aemelie Deelder

Spring Rain

Designer NOSIGNER

The Asian pasta "bean-starch vermicelli" is called "spring rain" in Japanese, named after the chinese characters of it. It is a lighting made by bean-starch vermicelli. As a matter of course, it is edible when boiled.

Vermicelli is a naturally made biodegradable plastic and can be said to be an ecological product for recycling use.

I propose "edible" products that leaves no trashes when dumped, for saving our living environment!

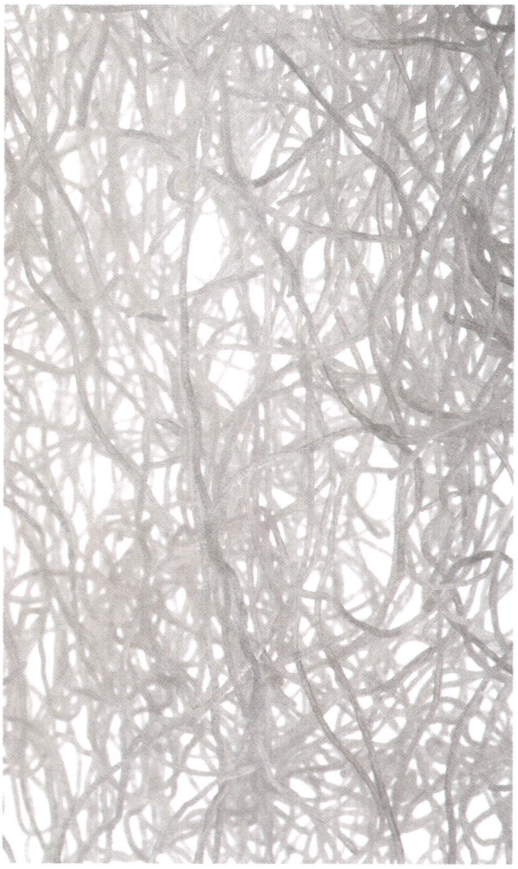

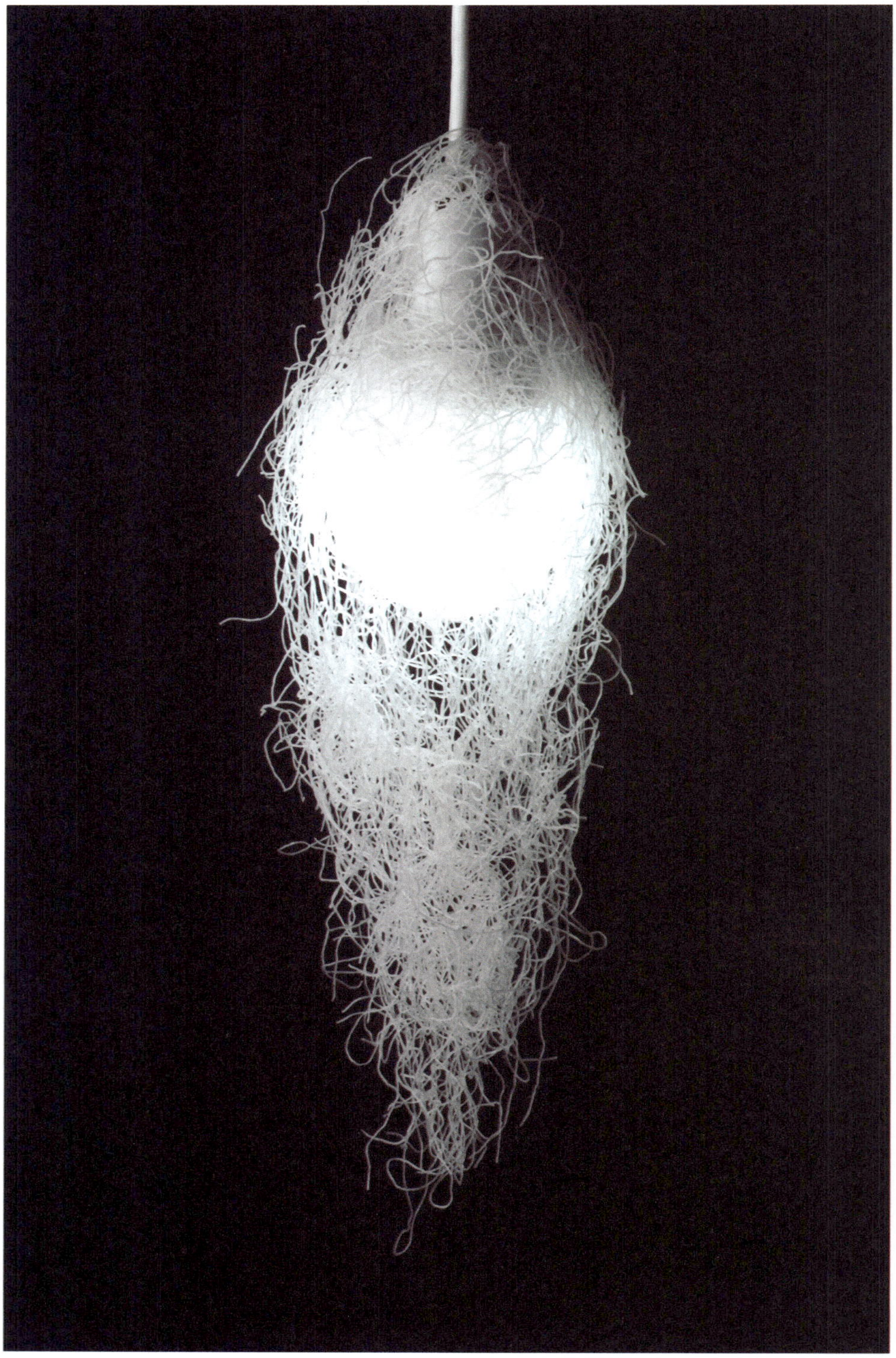

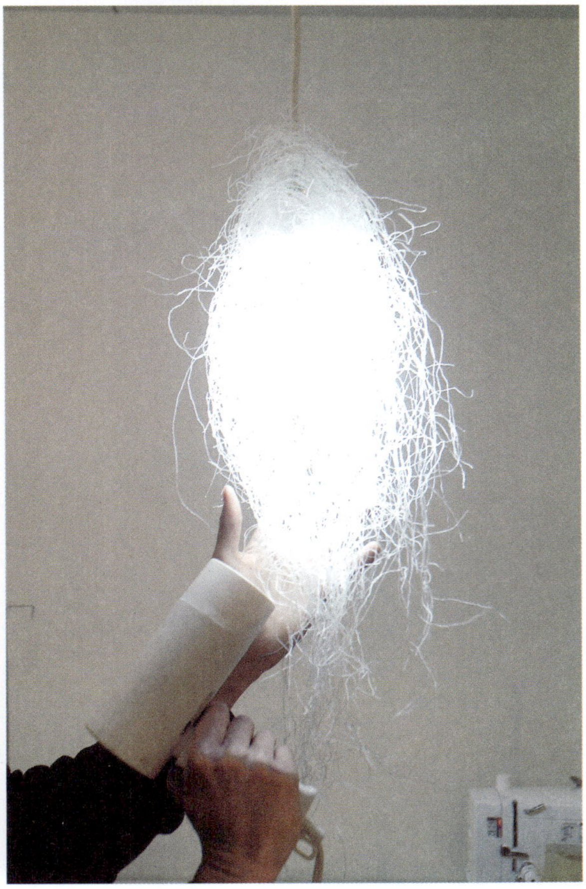

Rebirth

Designer NOSIGNER

Rebirth is a lighting product made by real egg shells.

Egg shells are a very fragile and weak material, but you can make a big structure by attaching them in a mutually complementary form. Now, the structure becomes an autonomously-controlled beautiful shape of necessity such as naturally-occurring shapes.

In this age with tons of artificial things around us, I feel many people desire to have "real things", which provide them an actual feeling of being, by experiencing nature or life itself.

This light, named Rebirth, is a design product which was born from life itself. When you realize the traces of lives on the surface of the egg shells, you may feel some special emotional ties between Rebirth and you.

Fishscape

Designer Aruliden (Johan Liden, Olivier Gregoire, Yifei Zha)
Photographer Dustin Ross (final images)

Fishscape is a reinterpretation of the classic fishbowl, transforming the basic home item into a beautiful new icon. Evoking the feeling of a mountain range, the Fishscape is made of hand-blown glass and manufactured by Gaia & Gino in Turkey. The textured interior holds up to 2.5 gallons of water.

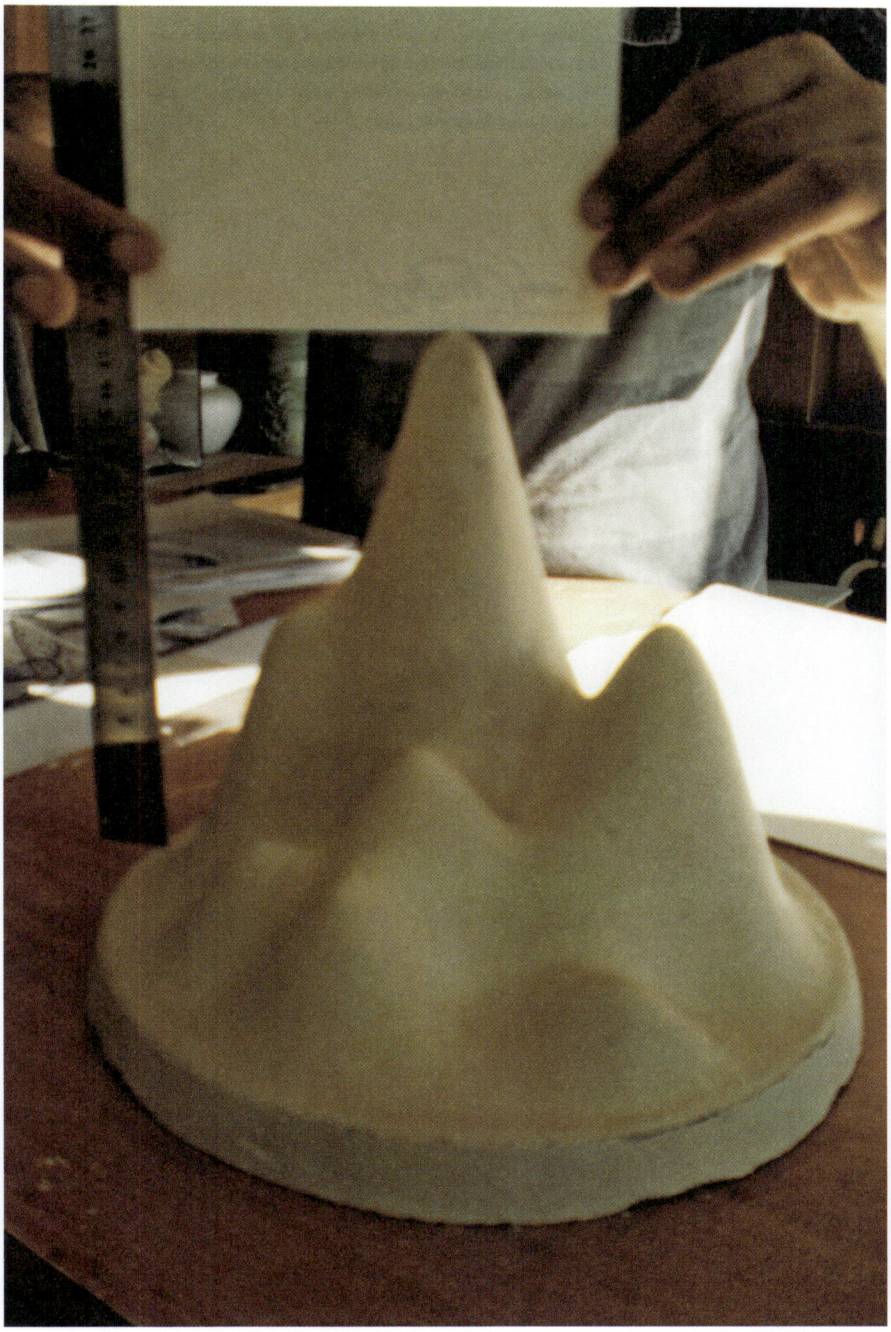

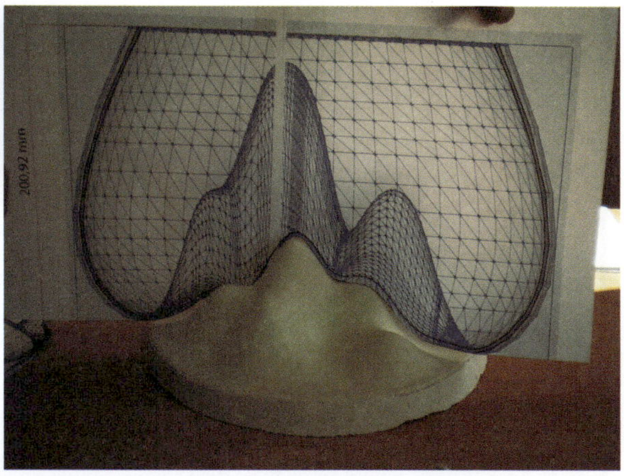
200.92 mm

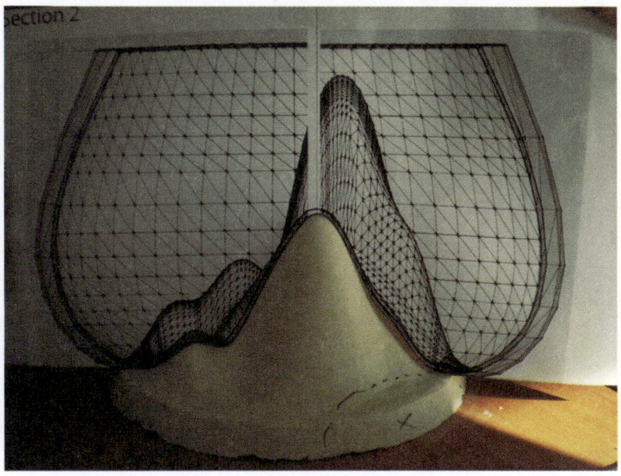
ection 2

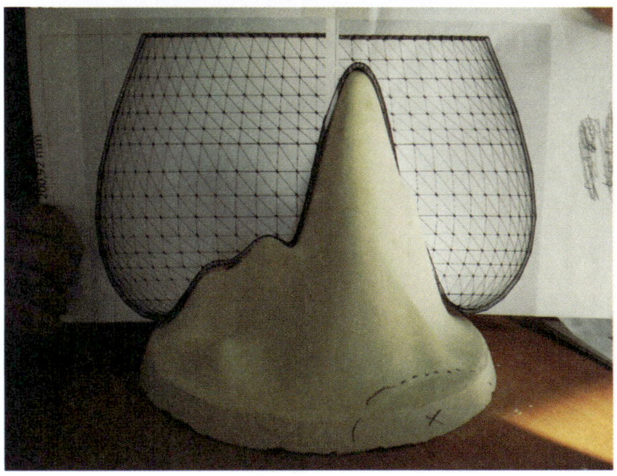

501 Chair

Designer   Assa Ashuach Studio

The 501 Chair form is designed to support sitting in half of a standing position. This position is the only way to achieve a free muscular tension sitting, where you can rest the back and lower lumbar areas while performing daily tasks like office work or a longer conversation...

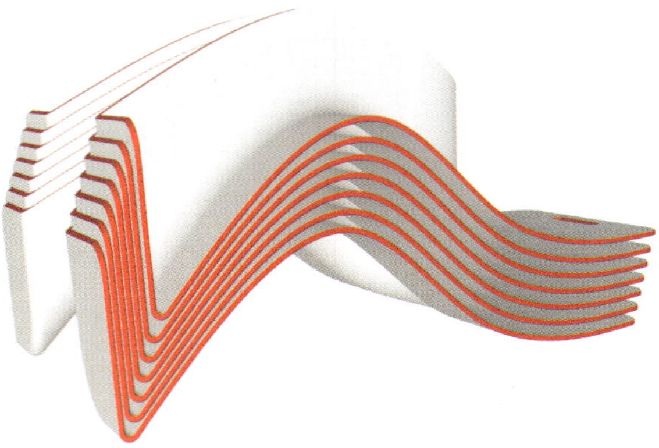

Upica Sofa

Designer Assa Ashuach Studio

The Upica Sofa was bought by private collectors and the British Council for their collections. The Upica Sofa was shown in many galleries and museums and was travelling through international venues in Europe and Asia.

The Upica Sofa design focuses on the tight relationship between the physical and the visual structure. This object was carefully designed to create structural strength while keeping the essential sculptural elements. The result is a thin and light structure, different from each angle when you look at it.

The Upica Sofa questions tradition. Technology today helps to simplify design. Alias Studio tools (3D software) give us the right tools to restrain and simplify a five meter-long line. The Upica Sofa is a composition of four lines that creates one compound surface. It is a very slim and large surface that supports itself. If a few millimetres of the surface tight shape were changed, the sofa would collapse like paper.

Fly Light

Designer Assa Ashuach Studio

Ai Light

Designer Assa Ashuach Studio

The Ai light has an artificial intelligence brain that senses space through five different sensors. Using a biologically inspired mechanism, this structure morphs and generates new behaviors according to your personal space.

When you first invite it into your home, you have to let it get accustomed to its new environment. Once it is relaxed, the training can then begin. It has five senses that track changes in its environment and slowly it develops a set of behaviors that indicate a new character to each light. The user is also able to interact with the light by playing with it through sounds, light and movements. This smart structure may behave in unpredictable ways if moved to an unfamiliar space.

One of the revolutionary aspects in this project is introducing the notion of "training" over "controlling" products. In this product there are no buttons and no remote control. The product will develop its behaviors by personal training and a natural interface.

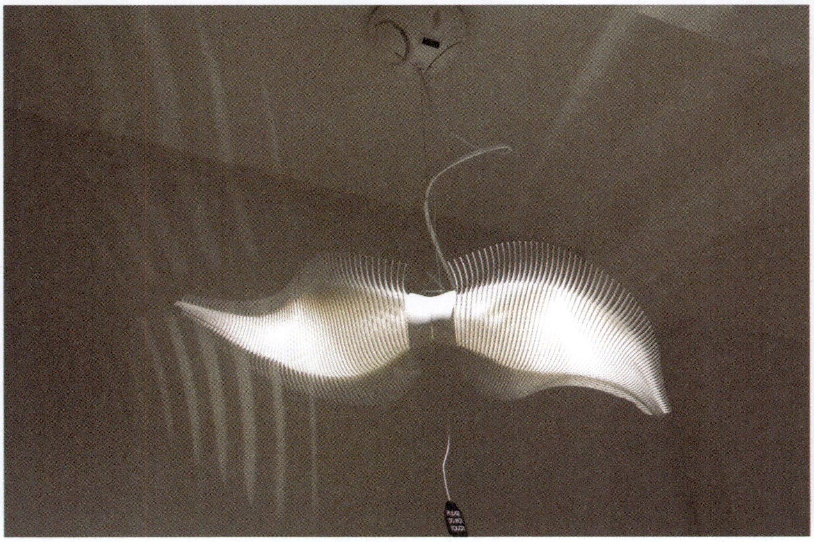

Loop Light

Designer Assa Ashuach Studio

Digital forming was first introduced to the public in an exhibition at London's Science Museum following a TSB grant in 2009. The notion of "open objects within boundaries" was introduced for the first time during Assa Ashuach's research at London Metropolitan University in 2005. During his fellowship, Assa introduced the Digital Forming® workflow. This novel workflow was created based on the assumption that 3D objects are fundamentally a line of code, a code that can be stored, embedded and rendered as a "virtual open product" (open, as with "no boundaries"). Online and offline, it can then be reconfigured and sent into production in the form of additive layer manufacturing.

My Dear Deer Friend

Designer Diploo Studio
Country Poland

Ceramic hand-made sculptures were designed for any interior bringing joy & excitement.

my
dear
DEER
friend
MEETS
MR
WOODS

diploo

Vaiss.eau

Designer Hanna Krüger
Photographer Minu Lee

The Vaiss.eau pendant lamp borrows from the forms of historical glass bottles and receptacles. The project is concerning the relationship of traditional and manual methods of production and the design of contemporary products. The lamps particular exterior form and especially the inward curvature at the bottom resemble the shape of old glasbottles. It has a mouth-blown double-walled glass body, which combines in a single form lampshade, reflector and illuminated space. The cable is coiled inside the glass body and technical details are visible: the image alternates between luminaire and technical glass apparatus. As soon as the object is filled with light, a vitreous light container seems to arise.

Brooklyn Bridge Lampshade

Design agency Judy Holme
Designer Judith Langdon

This contemporary lampshade designed by Judy Holme features the iconic Brooklyn Bridge landmark. It has been digitally reproduced from her original photograph and enhanced by using tints of black, red and olive green wrapped around the shade giving it retro appeal.

"Questions" Silk Cushion

Design agency Judy Holme
Designer Judith Langdon

At first glance these cushions appear to present a tunnel effect within a circular pattern, but on closer inspection, they have deeper meaning. The pattern is created from a string of questions using the words themselves as the structure for the design. The questions are a combination of everyday ("What day is it?") and thoughtful ("Can we have everything?").

Catch the Zero

Designer Yoshio Hasegawa
Photographer Maya Tatsukawa

The two dozens of soda cans are bundled with metal wire. The side surface of the can is made out of corrugated paper and the illustration board is used for the top and bottom. The white surface is painted with acrylic.

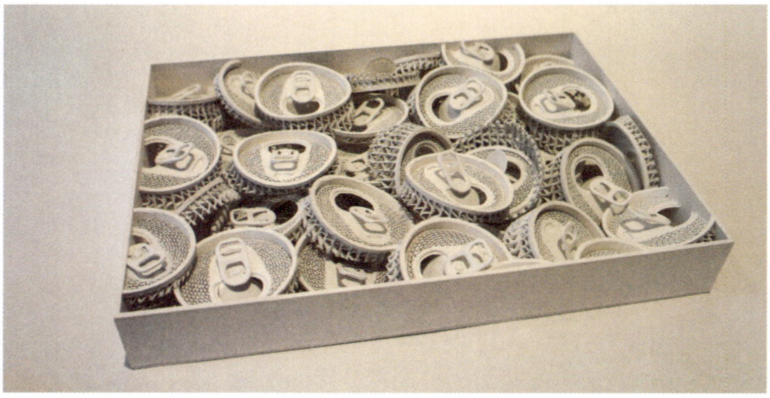

Warm Wind

Design Agency MiMo Digital Company Limited
Art Director Gary Yu
Designer Gary Yu
Photographer Mark

Nowadays, iPhone become more and more popular. The perfect shape makes a lot of iPhone user persue the iPhone case which closer to their habits.

As a result, our designers designed a novel and lovely iPhone case which is combining the plush toys and iPhone case together. And make this unique idea become real. This iPhone case make iphone user feel warm even in winter and win the good response in the market.

In the future, our company will design more produsts to customer.

Swan Vase

Designer Kenyon Yeh Design Studio

London-based designer Kenyon Yeh's latest project is this Swan Vase, created by making a mold from a salvaged chair leg, then poured in coloured resin.

Its slender, curving shape looks just like the neck of a swan. Of course, the surprising fact about the Swan Vase is that it wasn't created based of a real swan's neck, nor was it fashioned simply from the memory of what a swan looks like. The minimal curves and clean lines of the chair leg intended as a vase, once cast in resin, seemed to have transformed into an abstract fluid form - which makes a rather elegant ornament indeed. Finished off with a gilt 23K gold edge around the rim and a rectangle stand, the makeover's complete.

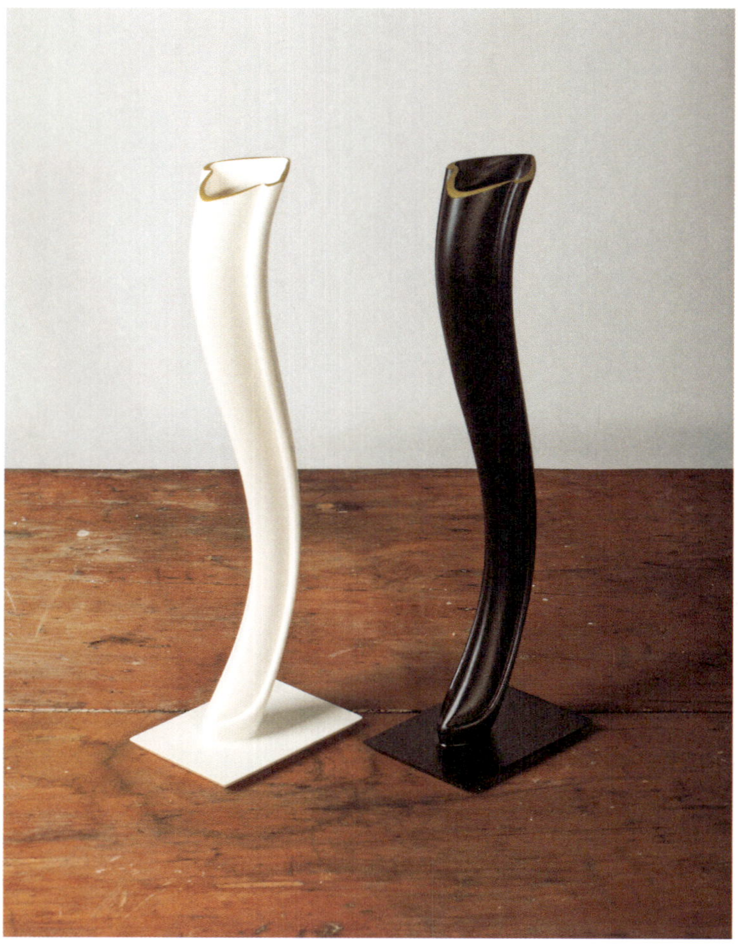

Stick Insects

Designer Kyle Bean
Photographer Owen Silverwood

A series of insects are carefully constructed entirely from matchsticks.

The Future of Books

Designer Kyle Bean

This Book/Laptop hybrid model is a response to the way we are sourcing information. Since the birth of the internet, we are relying much more heavily on the knowledge we obtained through websites, rather than published books. Kyle Bean sculpted the model almost entirely from a second hand hardback book. When closed, my model looks like any regular book, with the title "The Future of Books" stamped onto the spine. When opened, the book illuminates from a cut away panel, and a model keyboard is visible on the other page.

Pencil Shaving Portraits

Designer Kyle Bean
Photographer Victoria Ling

Kyle was commissioned by Wallpaper* Magazine to illustrate the contributors page for their latest "Handmade" Issue. Often using everyday and waste materials in both commercial and personal work, Kyle decided that it was quite fitting for the portraits of the handmade issue to be made from pencil shavings. He assigned a specific coloured pencil to each of the portraits and then proceeded to sharpen them. With tweezers, he then arranged the shavings to form the faces using photographs as reference. No glue was used, only careful positioning before they were photographed, so no one was allowed to sneeze or breath too heavily!

Rose Vase

Designer Louie Rigano

There is a perpetual human need to preserve the fleeting spectacle before it passes. Rose Vase is a study in both capturing ephemeral sentiments, and in vessels as well as the contents they hold.

The rose itself is an icon of fading beauty. Flowers have a self-destructive tendency: as they age, they wither and lose their petals. Here, this play of transience is hardened, the temporary is made permanent.

The vase is made entirely of rose petals, the remnants of the contents. Together, the vase and rose become part of an enclosed system, a dialogue. The outcome is a vessel that questions its contents, and a material that questions its origin and application.

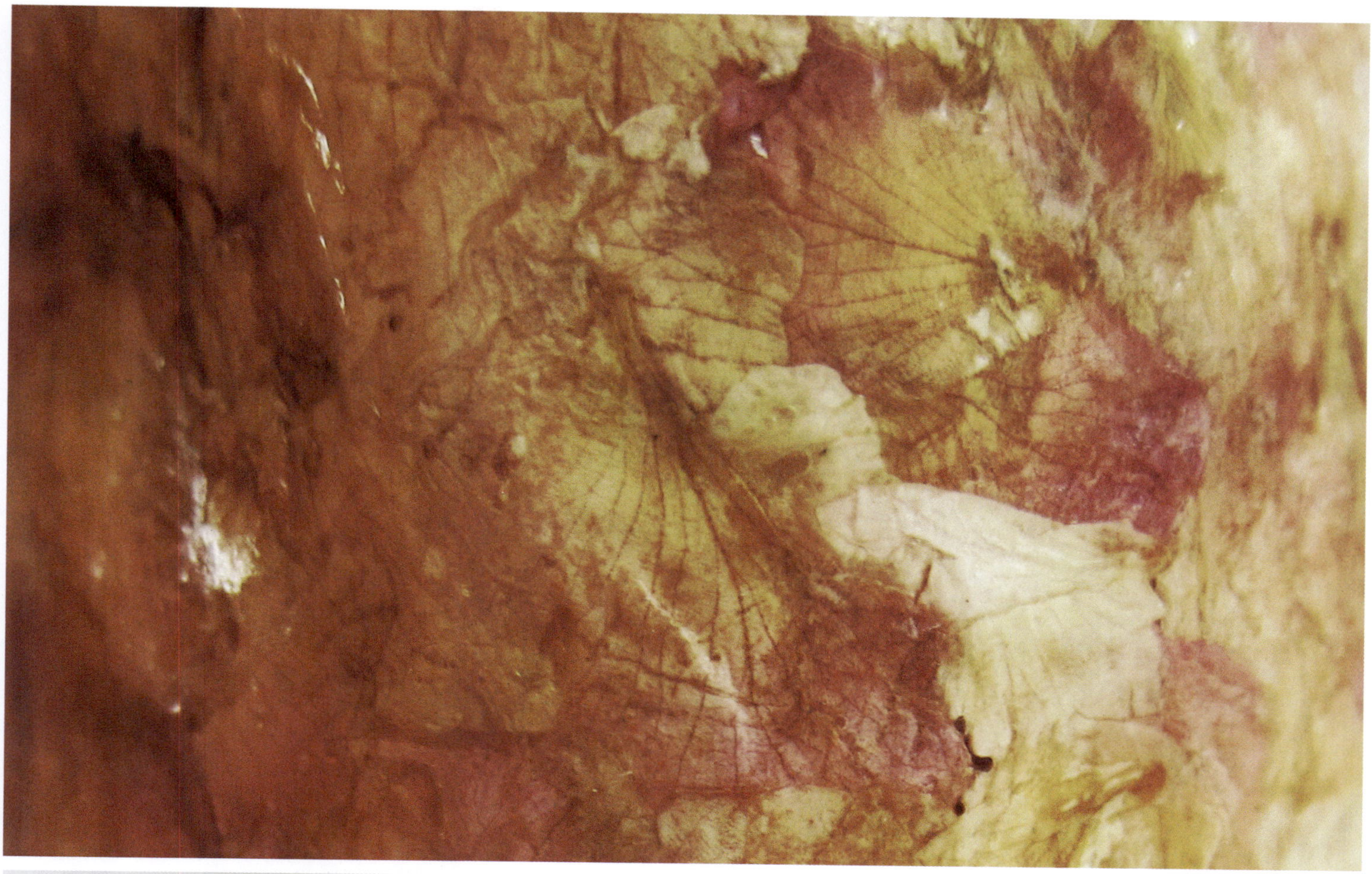

Teapot/Cup

Designer Louie Rigano

A series of teapots questioning perfection as an end-goal and exploring the relationships between textures and between mass-production and handmade.

The parts of the teapot that one comes into contact with are streamlined, smooth, and easily mass-produced. The body, however, is handmade and wheel-thrown, unglazed and rough. The unglazed ceramic body is easily capable of becoming well-seasoned after repeated use, which is a prized quality found in old teapots. The wooden fixtures, which get handled, are able to develop a richly aged surface and patina after repeated use.

The lids, which also serve as teacups, and the handles, both made of oak, are a standard size and shape and can be produced in multiples in industry with ease. The ceramic body is thrown by a potter who simply make the rim and handle plug the same standard size. Besides these two requirements, the potter has completed artistic freedom to create the teapot in any shape or dimension.

36 Pencil Bowl

Designer Michiel Cornelissen

There are different ways to look at this design. First of all, it's a tribute to the ubiquitous, and lovely, hexagonal colour pencil. It's also a representative of a new internet-friendly production model: combining relatively simple, locally available materials with high tech, custom designed interface parts.

The central connector part of "36 Pencil Bowl" is 3-d printed in laser-sintered polyamide, and can be combined with any set of 36 standard hexagonal pencils, available from art supply stores and department stores around the world.

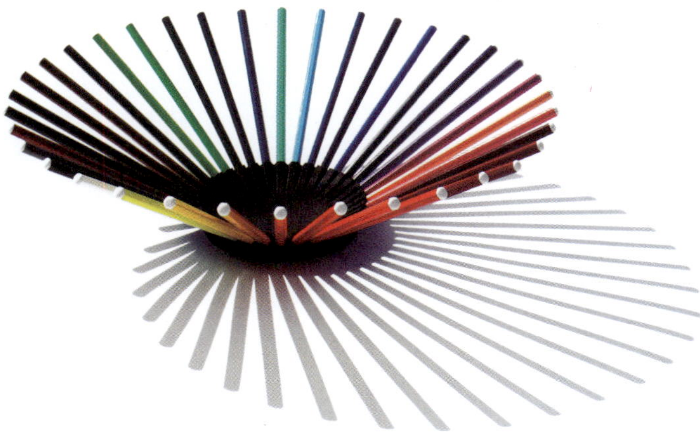

A Book Shelf

Designer Miriam Aust
Photographer Sabine Bielmeier

In nearly every household they are to be found: old books not read for quite a long time. Books are mostly presented in a shelf like trophies or remembrances although they probably will never be read again.

This book shelf offers these books which became precious just a new presence and aditional function in the special shelf architecture. A quite arbitrary number of shelf boxes in different heights, cutting and turning round each, are forming this shelf. Every crate is oriented in its height and depth on a pair of books and forms an individual frame for it. Put into grooves, with an extended lid each, the books are forming small hinged storage spaces in the shelf. Especially for the construction of such a shelf, a program was created which chooses the books for the individual shelves, arranging the boxes and drawing up plans.

Deliberately choosen was the contrast between dealing with used, well-thumbed books and the systematic specification for the creation of the shelf, allowing to produce, but with many individual pieces.

Hermann Löns
Sämtliche Werke
Hermann Löns
Sämtliche Werke

NATIONAL GEOGRAPHIC
Arundhati Roy
Der Gott der kleinen Dinge
Haruki Murakami
Mister Aufziehvogel

Vase & Leuchte

Designer Miriam Aust
Photographer Jan Köhler

At least one room plant can be found in nearly every living room mostly situated on the window-sill, but normally not integrated in the furnishings. The relation between human being and plant in a room is in the focus of this examination new ways to combine closer the living and plant cultures. The basic contents is the need of having the nature at home by means of a plant. In most cases a plant is cultivated in a flowerpot, its stature is controlled and the space permitted. The draft is sounding out possible barriers and analysing how much a plant can be influenced, but still remain "natural".

A series of 3 glass plant lights is the result of exceptional combinations and unusual things. The nature is presented artifcially, the living being used as material and laboratory aesthetics is brought into the living room. The basis of the draft is as well the unusual combination of two articles of different purpose but similar placement in the room. A lamp normally is installed in the ceiling above the dining room table, where in most cases a vase is placed too. Therefore the idea came up to combine both. The light is pulled down to the table, surrounded by a kind of cylinder, which can be filled with water and decorated with water plants or cuttings. Especially suitable are floating plants for fitting the glass body or cuttings of standard room plants. The 3 objects varying in their heighth just as roots of different plants.

The roots form the structure of the lamp shade in which the rays break. So the plant is not only decoration, but also a component of the object. The observer shall be spellbounded between the beauty of the light gleam broken by the naked water roots of the cuttings and the question whether a plant is treated right in such a sterile display. The 3 lights are consisting of glass bodies formed by an inner flask and an outer cylinder flask and cylinder are melted together at the bottom, so that the whole glass body looks like pulled inwards. The luminous means is simply fitted into the flask in using a stopper. For cleaning purposes, the luminous means can be simply put out.

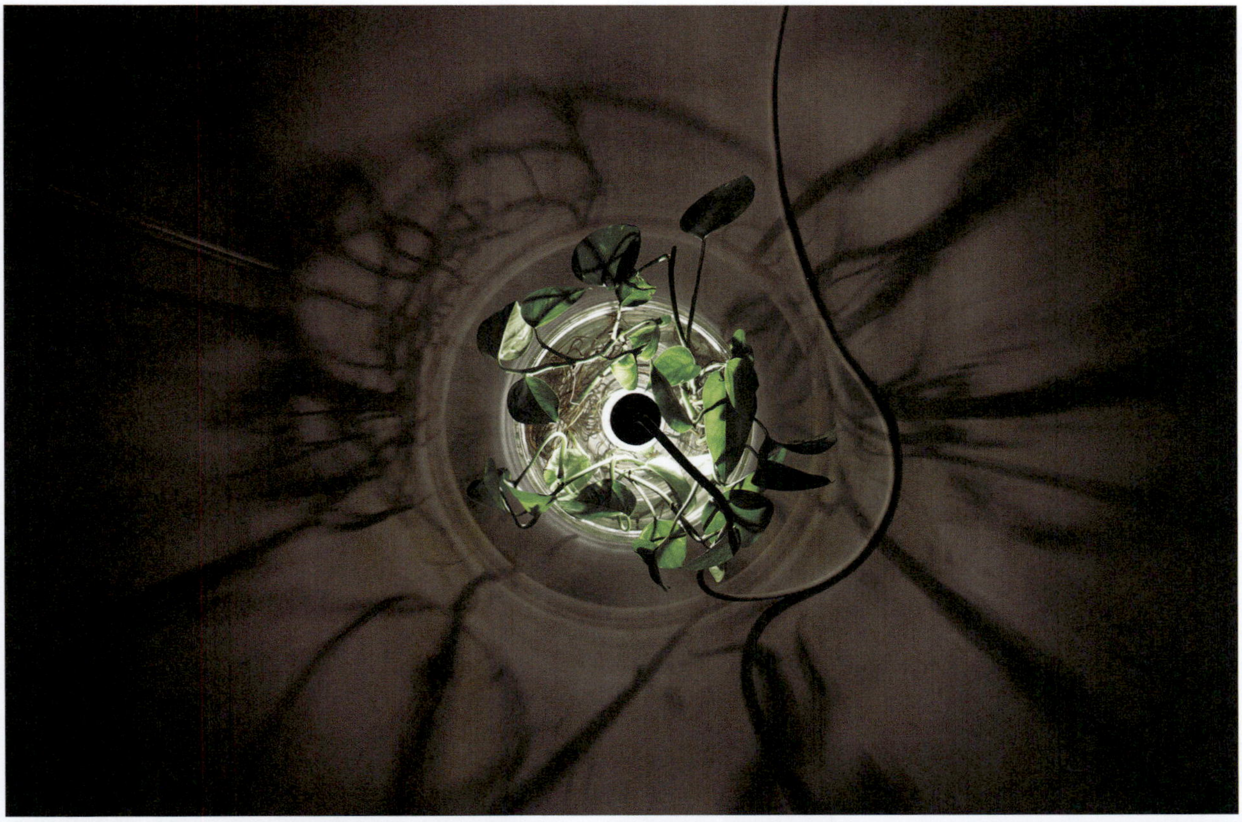

Wohnbeet

Designer Miriam Aust

By this special kind of bed a piece of garden is transported inside the living-place. Inside digging and weeding out is now possible and after having finished the work, one can sit there between the created splendour with muddy fingernails. The living and plant cultures are combined closer.

The bed is surrounding the objects of living and including these in its construction and creating a special living scene. The scaffolding-like furniture and objects of the home bed in their function are serving not only the people, but also the plant. The socreated home bed is only the basis for the process. The proper designing is handed over and done by the gardener and the plant.

Confectionery Bowl

Designer Oriental Fusion Handcraft[SM] – Recycling Workshop

Submitted in 2008 by Leticia Soto (Chicago, IL) as part of our "Crafts for Holidays Design Competition", this bowl was created from 3 half gallon milk cartons and decorative paper. Although this useful bowl did not receive any popularity vote award during the Family Day Festival in Chicago, many recycling craft members wants to make it. From the overwhelming response from the members, it received the honor of being including in our manual. Such a simplistic durable design can be used many times to entertain your holiday party guests with sweets, chips, or it can even be used as a centerpiece for a holiday dinner table or maintained daily as a fruits basket. The entire surface can be covered with paper to have a decorated look even if it remains empty. No matter what it is being used for, the milk carton inner lining makes this durable. With its popularity amongst members this bowl is popping up everywhere, at the office with supplies, at the parents meeting with treats, and even inside the classroom to hold crayons, markers, and toys.

Spiral Staircase Container

Designer Oriental Fusion Handcraft℠ – Recycling Workshop

The winner of the Most Creative Award 2010 for our "Desktop Organizer Design Competition", Yukie Partos (South Pasadena, CA) swept away the competition by adding its brand-new innovative design that is never before seen in our craft workshop. From the outside, this craft retains the full original shape of a half gallon milk carton, but because of the new vertical hinge concept and its spliced segmented center, it swings out to reveal multiple layers of useable containers with the help of a second half gallon container to make bottom trays. The wooden stick used for the hinge must be smooth; however it could be created with a long unsharpened pencil. The inspiration taken from a spiral staircase breathes life into the traditional office desk organizer transforming the old milk carton, which once would have been trash, into a work of art. Not only did Reiyukai America get a lot of comments around the office, but also found a lot of discussion from both Mexico and Japan on their Facebook page. This beautiful piece brought a lot of people together and inspired many to become new recycling hobbyist. The designer also left the carton cap on top so everyone can tell that it was made from milk cartons!

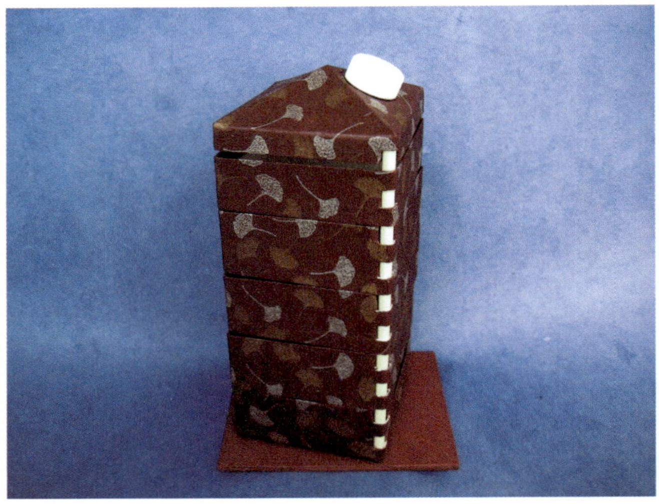

Around the Clock

Designer Peleg Design

Around the clock is a cylindrical desk clock made from aluminum. A new aesthetic is achieved by tilting the cylinder in an angle and bending the dials.

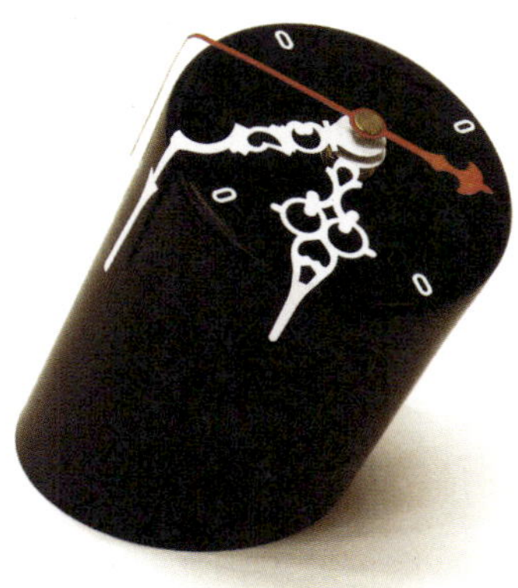

Pegs Memo

Designer Peleg Design
Photographer Studio Dan Lev

This memo board is perfect for your important reminders, like coming up birthdays, unpaid bills or anything we tend to forget.

The mirrored finish stainless steel board reflects half of the magnetic pegs and completes them to appear as if they were a whole peg.

XO

Designer Peleg Design
Photographer Studio Dan Lev

A reflective Tic-Tac-Toe set that uses the reflective properties of the mirror like game board to create the illusion of playing with full pawns.

Reflective-Bookend

Designer	Peleg Design
Photographer	Studio Dan Lev

A set of 2 bookends is made of polished stainless steel. The reflective bookends illusion is a clever use of a mirror reflective properties to create the word BOOK on one end of the bookend and a exclamation mark an the other end, when viewed at an angle.

Magnetic Dots

Designer Peleg Design
Photographer Studio Dan Lev

Colourful magnetic stickers with built in super strong magnets will elegantly hold small metal items. Ideal for holding on smooth surfaces in the kitchen, shower, study or anywhere else in the home or office.

Baboo

Designer Shahaf Ben Abu
Country Israel
Photographer Meital Eshed

Baboo is a mobile kit containing wooden clips designed in different shapes. The clips are holding the children's drawings, and bringing them to life. The box is designed as a puppet theatre stage, this way it can be used in Drama Therapy sessions and as a diagnostic and therapeutic tool.

The clips, combined with the children's drawings, allow a rich and unique expression of the child, and can be played with everything according to need. I was born in Sderot, a town near Gaza that has been bombed by terrorist organizations since 2002. The project was motivated from the need to treat the children of Sderot, who are suffering from post-traumatic symptoms in the past few years. The research for this project was done in collaboration with Sderot's children.

This kit can be used as a tool for treating children who suffer from trauma, but can also be used as a regular game for fun and entertainment.

Baboo

Baboo

Mistral

Designer Spell

The Mistral pendant lamp, inspired to bright up the path of Little Red Riding Hood through the dark forest. A playful and mysterious light that is eye catching and brights up every dark space.

Piggy Bank

Designer Spell

The Piggy Bank stands as a metaphor for desire. Throughout the ages pigs have been slaughtered for feasts in honor of guests. The Piggy Bank engraved lines translate the value of each cut and remind the joy of saving to make your dreams come true.

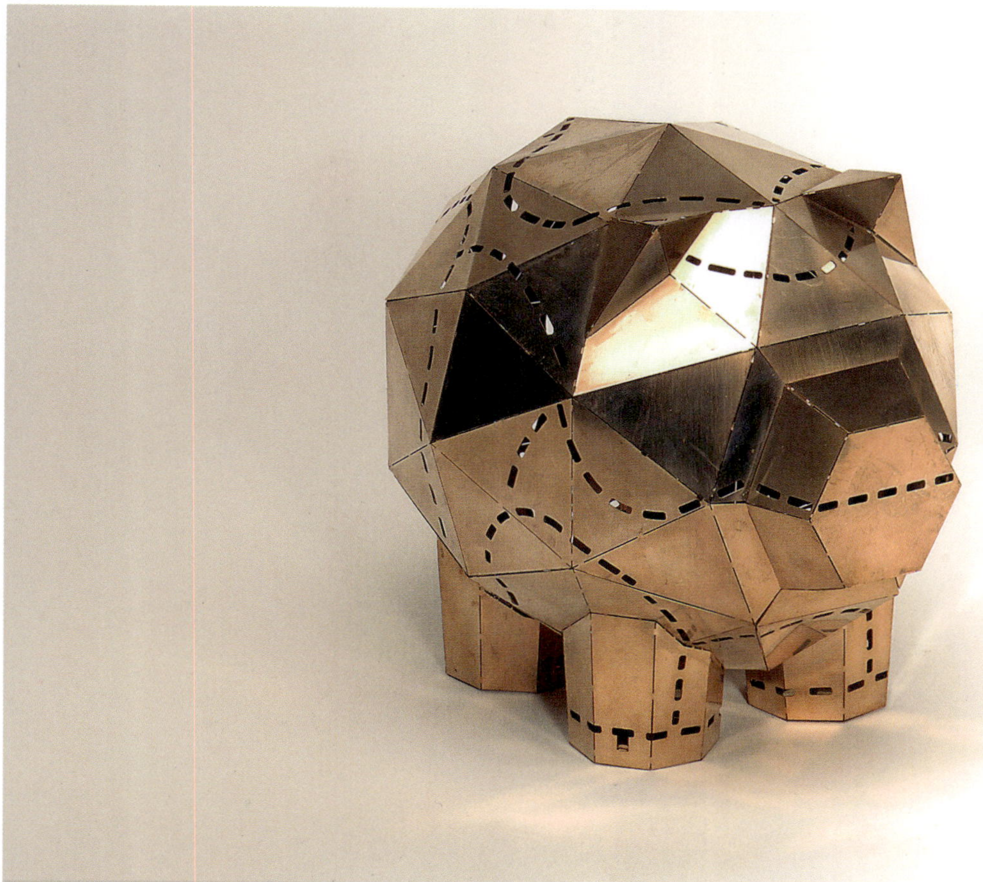

Nomad

Designer Spell

Living and working environments become more dynamic. Where relaxing and working places overlap, Nomad tablet table takes place! The various range of tablet systems can be used for waiting, working, relaxing or even for a small meeting.

Sweet Stool

Designer Spell

Like candy, Sweet is dressed in a layer, that is durable and delicious in sight and sitting comfort.
Available in a variety of colours, to become a cheerful addition to any interior.

Awaglass

Designer　　Studio Note

You can enjoy a time during filling it up in a hourglass by bubble.

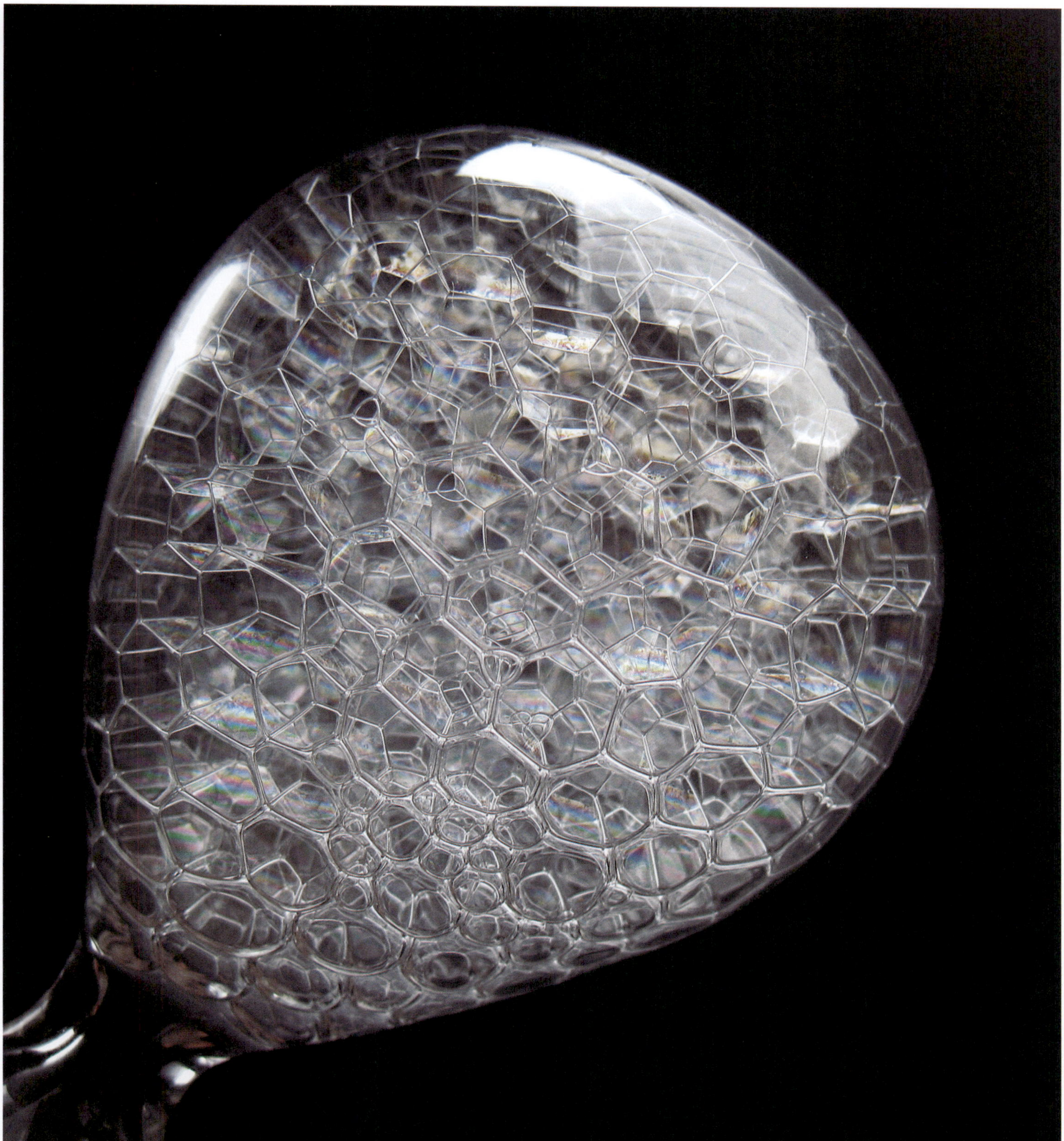

f,l,o,w,e,r,s

Designer Studio Note

When it goes to the field and the flower queues up at intervals of 1cm, the height can be measured by lying.

Glassmic Vase

Designer Studio Note

Sometime when I make a new product I have to break another product by using a part of it. This is like a child born from mother. I wanted to make a pruduct that has this story.

r,o,o,t

Designer Studio Note

The root is growing under the flower. Normally we are not able to see it but it would grows up at intervals of 1cm, so I made setsquare by enclosing in an acrylic.

Tile Cowpet

Designer Studio Note

The rug is a part of furniture. The tile is a part of building. I would like to make a product that have both elements.

Twotoo

Designer Studio Note

If two cups cracked to the same shape, I can remake new two cups by mixing.

Fuzz

Designer Study O Protable

The project utilises a void as the main tool of production. A geometric shape is made form wax and the ceramic resin is brushed on in layers, blurring the rigid shape into organic roughness. When cut open, the resulting object becomes a container. A container as an object has a specific function determined by the size and shape of the empty space (void). Fuzz become obsolete in function as it loses the ability to accommodate the original shape.

Glass Tube Necklaces

Designer Study O Protable

Glass Tube Necklaces are glass version of Drainpipe Necklace, with hand bent borosilicate glass tubes and sterling silver chain.

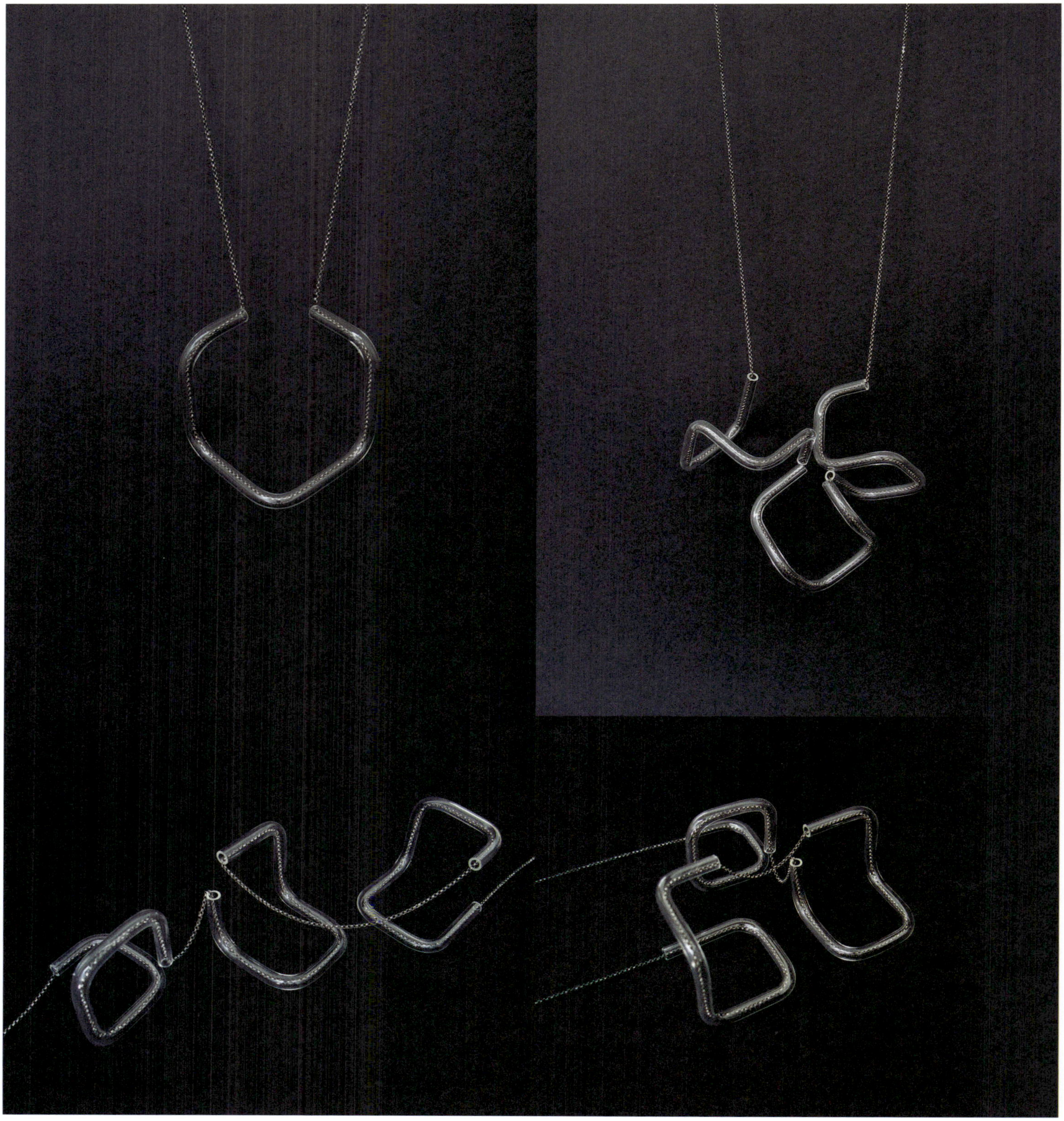

Leather Bangles

Designer Study O Protable

Rolled-up and riveted, hand-dyed leather bangles in a multitude of colours.

Neo

Designer Verena Klette

Ring and necklace

Transit

Designer Verena Klette

Pin

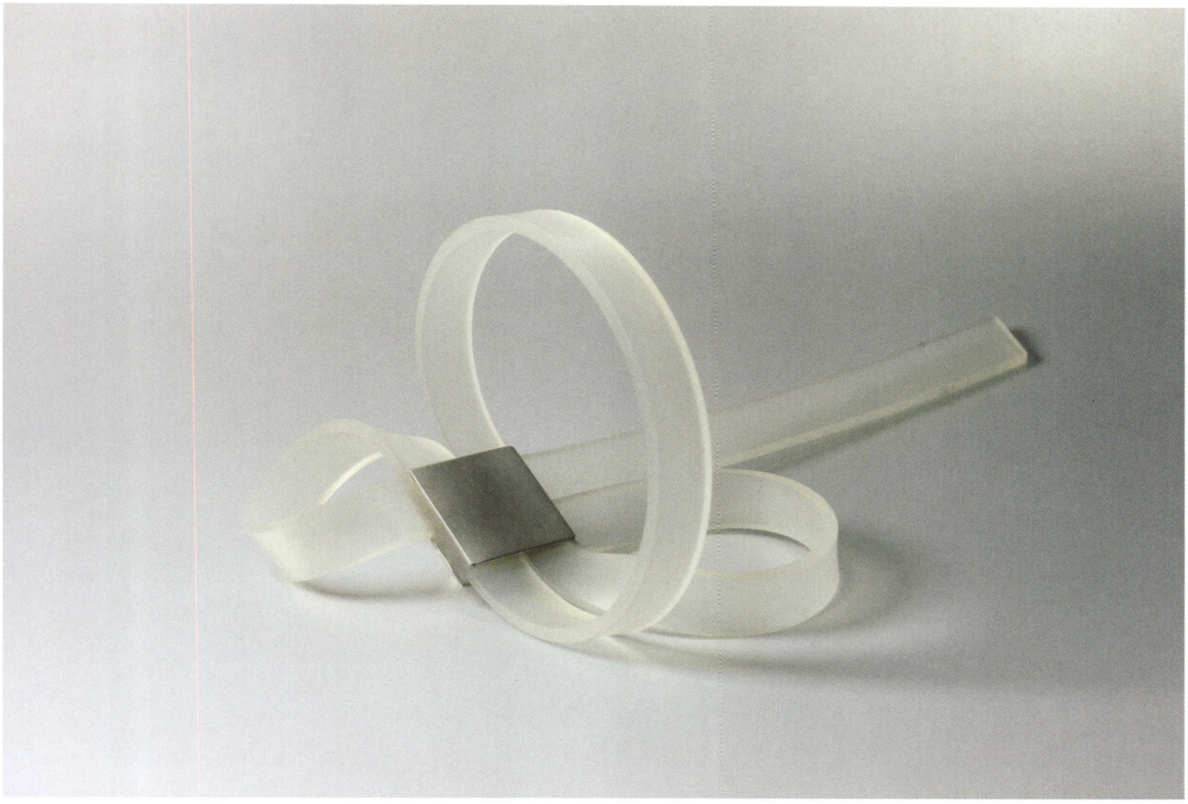

Reconfiguring the Ordinary

Designer Yong Joo Kim
Photographer Studio Munch, Affandi Setiawan and Yong Joo Kim

Reconfiguring the Ordinary: Stacked and Twisted
Velcro and Sterling Silver
3.5" x 3" x 3'
Brooch

Reconfiguring the Ordinary: Aligned, Rolled and Attached
Velcro and Sterling Silver
18" x 8" x 0.5"
Neckpiece

Reconfiguring the Ordinary: Rolled and Inserted
Velcro and Sterling Silver
1.25" x1.25" x1.25"
Ring

Reconfiguring the Ordinary: Rounded, Aligned and Folded
Velcro and Sterling Silver
3.5" x 4" x 0.5"
Brooch

Reconfiguring the Ordinary: Rounded, Alignedand Twisted #2
Velcro
3.75" x 4" x 1"
Bracelet

Reconfiguring the Ordinary: Twist Looped and Linked
Velcro
12" x 16" x 4"
Neckpiece

Reconfiguring the Ordinary: Rounded, Aligned and Twisted #1
Velcro and Thread
8"x 15" x 3"
Neckpiece

Reconfiguring the Ordinary: Twist Looped and Attached
Velcro
14" x 15" x 2.5"
Neckpiece

Fumfum

Designer Fumie Shibata
Client ADMIRAL SANGYO CO.; LTD

Flower magazine holder

"Flower" is a flower shaped magazine rack. It holds the magazines and newspapers standing on the desk and enables to put them in order, like flower decoration.

Solo vase

A single flower could be arranged with the figure in the natural bloom. A frog is built in the vase to hold the flower fest. The flower shows itself so fine, from the stalk to petals.

Nine Hours

Creative Director Fumie Shibata
Client Cubic Corporation.
Photographer Nacasa & Partners Inc.
Design Director Fumie Shibata
Graphic Design/Sign Design Masaaki Hiromura
Interior Designer Takaaki Nakamura

Shifting from the traditional, 9h/nine hours proposes an unprecedented idea of lodging service in an urban area.

The service represents "the space for sleeping in a metropolis".

9h is the time spent for three basic actions: take a shower, sleep, and get ready: one hour + seven hours + one hour. Based on this concept, 9h offers urban stay for people, on business for instance, who spend a busy and active day in a metropolitan area.

9h, an innovative commodity enriched by the unique design and hospitality, is about to be exported to numerous places in the world.

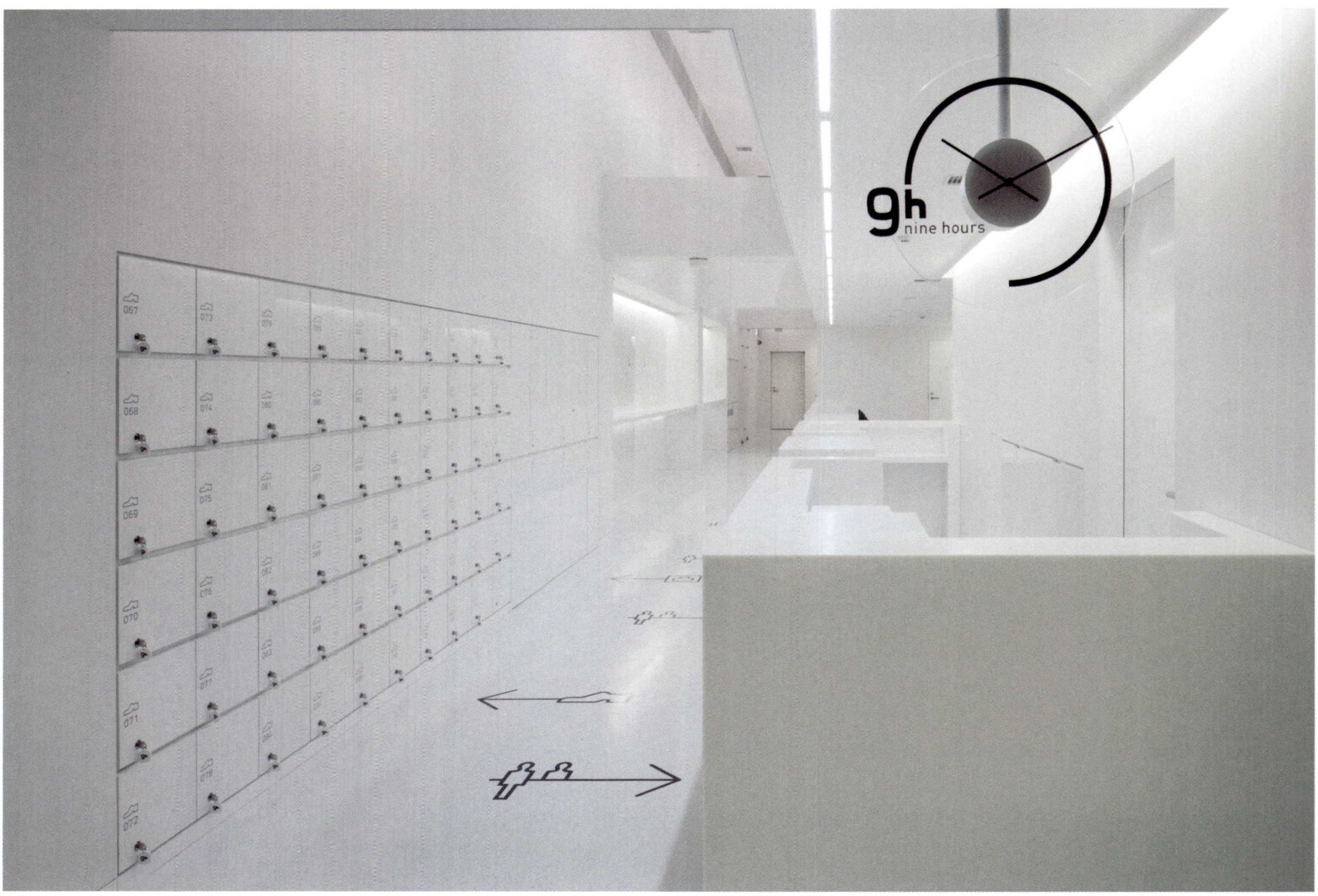

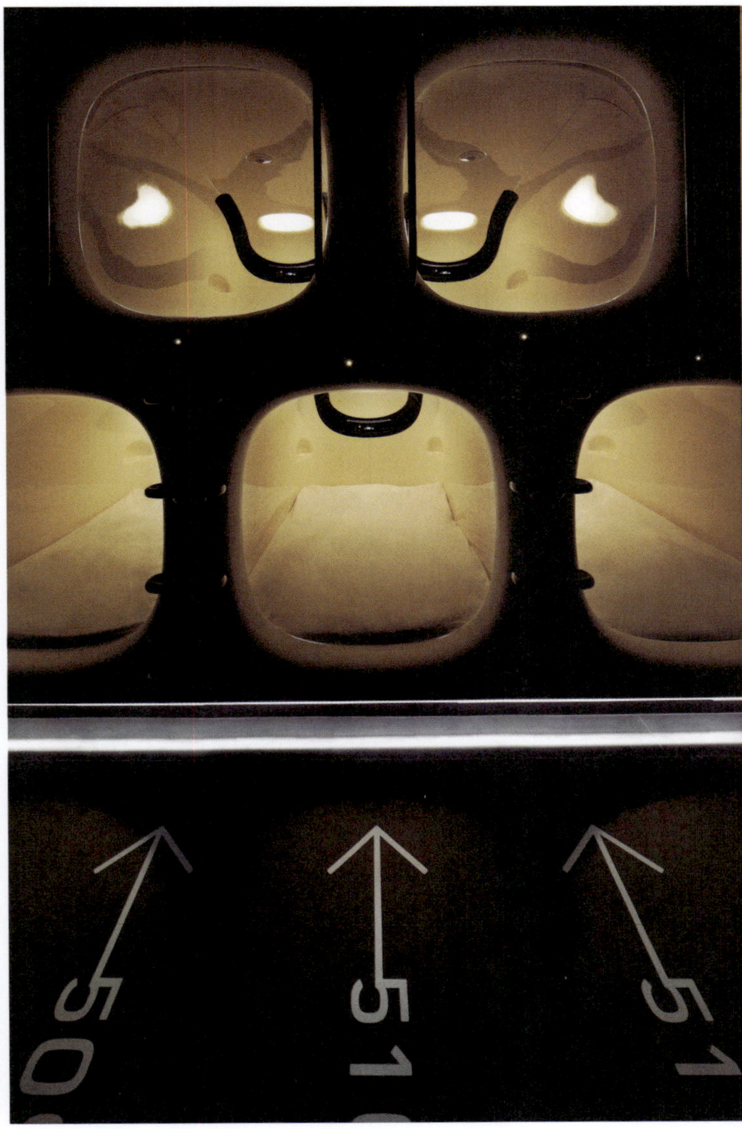

507
506
505
504
503

Back to Basics

Designer Zim & Zou
Photographer Zim & Zou

We've made those papercrafts for several reasons. At first, it's a tribute to vintage technologies which marked the technological evolution of the last years, and all the nostalgia of the memories that each has with them. By getting these "obsolete" objects back to life, we tried to highlight the very fast evolution of our everyday possessions. The devices we use nowadays will in a few years be considered relics too. We decided to use paper because more than the fact that it's a basic material, accessible for everyone, it's a way to show the balance of power between digital and paper production. When today you can read a book on a screen, we needed to create "real" things. The aspect of craftmanship is really important for us. We wanted to turn an industrial object into a unique handmade craft. And the fact that the objects are not working places the user into a spectator position, a way to see the object out of its function.

TETRIS
SONY
WALKMAN
POLAROID

New Tysons

Designer Zim & Zou
Photographer Zim & Zou

The story is about an area in the Washington DC suburbs, that is mainly offices, shopping and parking lots with tons and tons of cars. 4 metro stations will open out there. This car driven area will suddenly have more public transit. Right now it's kind of a soulless place, but they're hoping to make it more of a city. Hoping to add more pedestrian walkways, more apartments, more grocery stores - more of a place that people want to live, rather than just work.

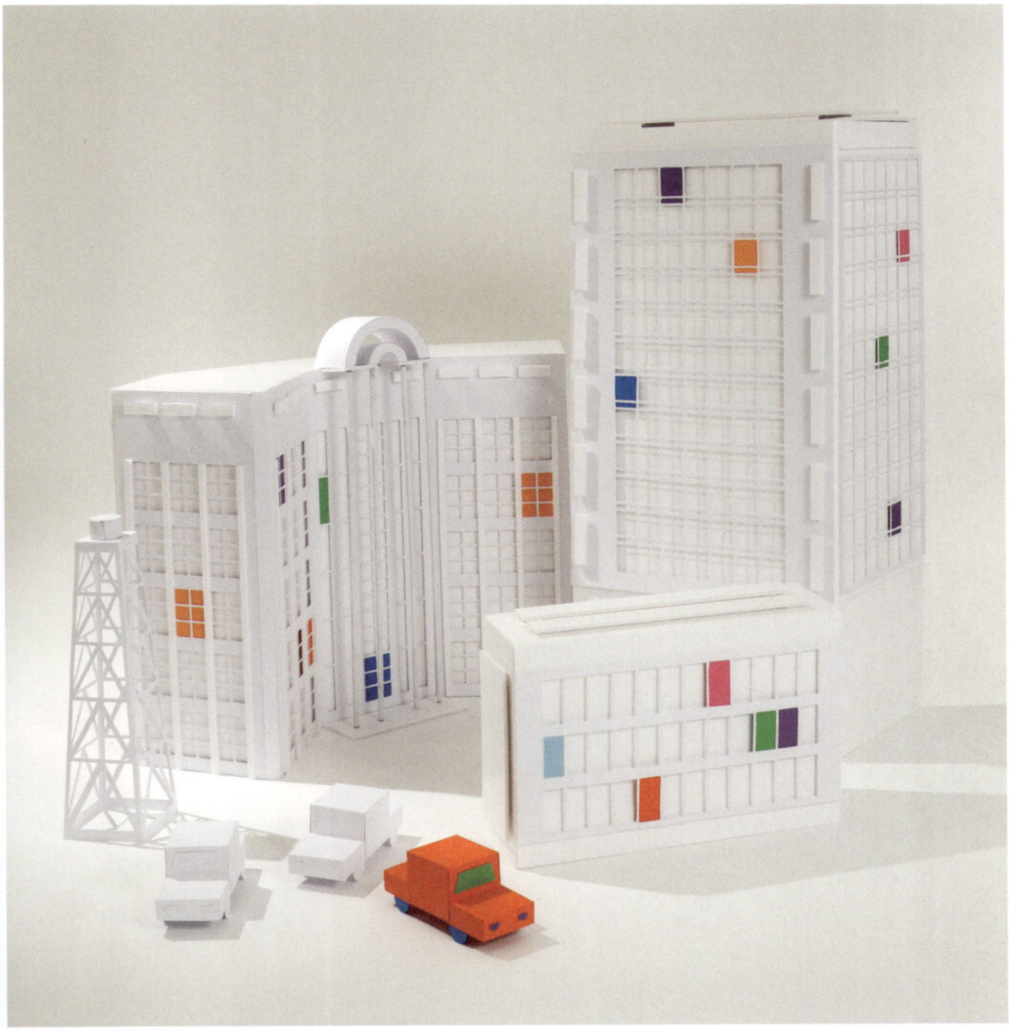

Weave Type

Designer Zim & Zou
Photographer Zim & Zou

Weave Type 2 is a handmade font made by thread and paper. It took about 6 hours per letter and 500 meters of thread.

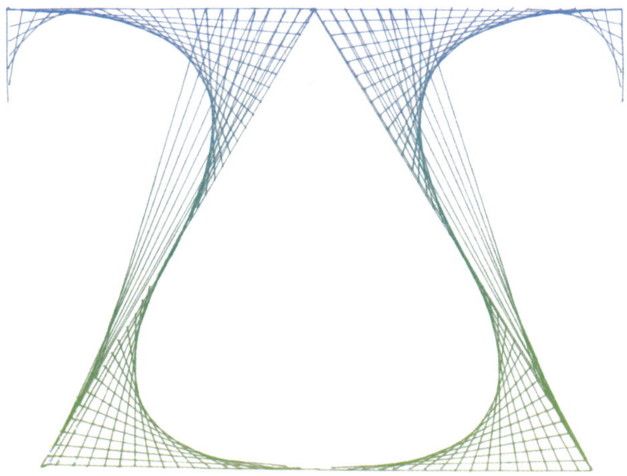

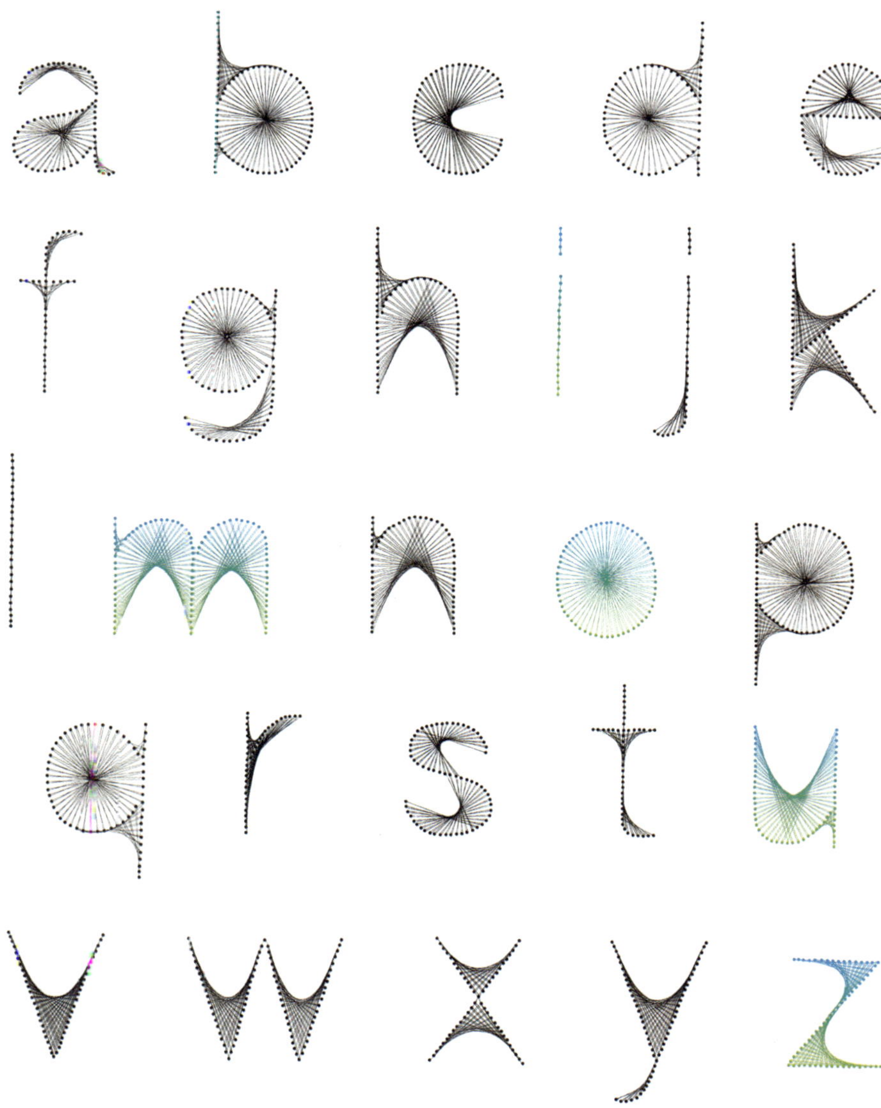

ABCDEFG
HIJKLMN
OPQRSTU
VWXYZ&

ABCDEFG
HIJKLMN
OPQRSTU
VWXYZ&

WE

AVE

Thanks

A

Aruliden

Aruliden is an award-winning product design and brand strategy consultancy based in New York City, founded by Rinat Aruh and Johan Liden in 2006. Bringing over 25 years of collective experience in the fashion, automotive, lifestyle, and beauty industries, they use design to rethink brands and drive business. Through their process of producting, aruliden has been internationally recognized for their work.

Assa Ashuach Studio

Assa has also Co-found Digital Forming and UCODO. The two London based companies are set to democratise the personalization of everyday products; introducing the notion of ODO-Original Designed Object and CODO-Co Designec Object, using a set of innovative 3D software solutions for online customization and shape modification of designed objects. Digital forming was awarded government funding and was invited to show at London's Science Museum's 100 years anniversary fecturing "the future of manufacturing" showing for the first time shape changing 3D modifiable products designed by Assa for UCODO.com.

Assa is currently managing his London-based studio. He is the creative director of Digital Forming and UCODO. Assa is leading the "MA Design Suite" of London Metropolitan University, Sir John Cass Art Media & Design and sits on the advisory board of London's Digital Manufacturing Centre, Metropolitan works.

D

Design nobis

Design nobis is a sustainable design centre. They introduce innovative solutions in product design, branding and space creation and help their clients to differentiate in market by creating products and services with strategic value. Handling design services from research and concept development phase through prototyping and production, they aim to become a sustainable innovation centre.

Design nobis contributes to national and global industry by developing competitive products with high added value and they help to create sustainable development locally and globally. they provide R&D services for firms and add value to brand image in order to turn technology and inventions into products, services and good that are available to the consumer in the market. Through their designs, they create value added products while at the same time differentiating brands.

DING3000

DING3000 is the studio of designer Carsten Schelling, Ralf Weber Sven Rudolph and husband. After mutual study and work in various design studios such as Vogt & Weizenegger and Marcel Wanders, they have worked together since 2005, and design with a small team of employees, such as consumer goods, furniture, lamps and home accessories. Becoming known, they are with designs that are characterized by subtle, charming ideas, such as PIMP MY BILLY (2005), a series of "tuning-articles" for "Billy" from Ikea, the best-selling rack system in the world, or S-XL CAKE (2007), a cake pan made of silicone, the means of built-in portioning sections and different levels of plateaus of different pieces of cake shapes.

With their designs, they have so far won over 25 design awards, including the internationally renowned iF Product Design Award, the Good Design Award, the Red Dot Award and the Design Plus Award. In addition to a variety of other publications at home and abroad, our work has been featured in several books, including "Design Now!" pockets of the publishing house, "Furnish" and "Desire - The Shape of Things to Come" by the Gestalten.

Work of DING3000 are represented in the permanent collections of the Badische Landesmuseum, Karlsruhe and the Kestner Museum in Hanover and have to design festivals and exhibitions such as shown in the German Pavilion at EXPO.

Diploo Studio

Diploo Studio stands for ideas, that are hidden in our imagination since our childhood. Everything is inspiring, especially our memmories. Brought to the "surface" of reality, they are being carefully embeded into forms of various materials such as ceramic clay & others...

F

Fumie Shibata

Fumie Shibata is a product designer based in Tokyo. Focusing on product design, she has expanded her design activities in industrial design from daily goods to electronics and healthcare products. Her works have received tremendous acclaim in various design awards worldwide including Gold Prize of if Award, red dot Design Award in Germany and the Good Design Award in Japan. A judge for the Good Design Award.

G

Gary Yu

As the seasons change, the fashion change. Plush doll and toys always play a cute and warm role in our life, especially in winter. Plush doll and toys was alway well-liked, especially In young generation. They tice style still very typical.

Giancarlo Zema Design Group

The Giancarlo Zema Design Group is an architectural studio in Rome, specialized in semisubmerged architectural structures, marine parks, floating habitats and yacht design. It was founded in 2001 by the architect Giancarlo Zema as a dynamic and flexible organization, which guarantees the whole design process, from the concept to the executive details. Their projects are frequently shown on TV programmes and published in the most important architecture, design and nautical, magazines and books in all over the world.

Giovanni Levanti

Giovanni Levanti was born in Palermo in 1956. After graduating in architecture he moved to Milan. His objects are present in the design collections of the Triennale di Milano, the Museum of Fine Arts in Montreal and the Centre Pompidou in Paris. Among the awards, the Design Plus Prize 2000 and selected the XIX ADI "Compasso d'Oro" Award.

Green furniture Sweden

Green Furniture Sweden has its roots in Berhin Studios, created by the designer Johan Berhin in 2005. Green Furniture Sweden represents the furniture collection of Berhin Studios.

Smart is green and green is smart. To be green is to use the wonderful materials handed to us by nature, and using them with finesse – sparingly, sustainably, beautifully... Green Furniture Sweden thinks green, from the concept stage to recycling and regrowth.

gt2P

gt2P was born when Alexy, Guillermo, Tamara and Sebastian found that they had shared interests in the productive processes of architecture and design, such as manufacture, construction, and the possibility of generating them with digital media.

The concept of computer aided design knowledge has provided support to this studio, so it has moved forward to the innovation in design methodologies, as well as the integration of processes and the massive customization of projects. Otherwise, it offers a platform of permanent investigation in new materials, processes and designs, giving passage to the incorporation of methodologies that allow the development of digital crafting.

Under the concept of "standard methodologies for non-standard solutions", gt2P has achieved the aim of personalizing each one of its products and projects.

The development of products from the approach of the parametric design and the digital manufacture has taken gt2P to be selected to participate in Greenhouse at Stockholm Furniture Fair, SaloneSatellite in Milan, 100% Futures at 100% Design Londres during 2011.

H

Hanna Krüger

The Design Studio Hanna Krüger is working in the field of experimental design, product design and interior design.

In the design of objects Hanna Krüger is interested in the simultaneous coexistence of the raw material and the resulting new which comes out of it. One important theme in her work is the moment of production, in which the material, human, production methods and cultural context are converging. This moment has to be choreographed in every project.

I

I.D. Monica Navarro Aranda

Monica Navarro is a young industrial designer from the Industrial University of Santander, Colombia. She graduated with an outstanding thesis, trap for collection and surveillance of the mosquito Aedes aegypti, that was chosen to be patented by the university. She has projects in the fields of product design, urban design, assistive technology and research, and also has worked with CD & I Associates Colombia.

J

Judy Holme

Judy Holme is a British design label within the accessories market. The style is very graphic – incorporating a variety of media and the resulting effect is an eclectic mix of images. Judy uses the unique combination of expressive graphic design with digital print technology, which results in a new approach to surface design in both the home and fashion markets. It stems from a desire to break out of the predictable mould of patterns and motifs surrounding textile design with a particular interest in using words as a means to convey a visual message. With an instinctive sense of spatial layout and individuality, the use of photographer, calligraphy, block colour and illustration typifies Judy's work and attention to detail is often a key element in the design process. Judy takes sketchbook ideas and transforms them into beautiful stylish designs incorporating many themes from urban and poetic to classical and iconic.

K

Kenyon Yeh Design Studio

London based designer Kenyon Yeh was born in 1983, he graduated from TungHai University, Taiwan, as Bachelor of Industrial Design in 2007. Since he continues his study to London, he studied Master of Arts in Design at Kingston University, London, under Colin Holden and Simon Hasan and graduated in 2009. Through his passion in designing furniture, lighting, and product, in late 2010 he starts his own design studio based in London.

Kenyon has his own inspiration of designing his collection, where he believed that personality and characteristic can be transferred into object creating.

Kyle Bean

Kyle Bean is a Brighton based artist and designer specialising in hand made models, sets and tactile illustrations. Since graduating from a degree in Illustration from the University of Brighton in 2009 he has worked for a diverse range of clients on various projects. Kyle has created window displays for luxury brands such as Liberty, Selfridges and Hermes and editorial commissions for publications such as Wallpaper*, Financial Times, VMAN and Wired. Alongside his commercial work, Kyle has produced some personal projects that have gained him international recognition. These projects include "Mobile Evolution" - a Russian doll style model depicting the miniaturisation of mobile technology, and "What Came First?" - a sculpture of a chicken made from egg shells. Kyle has exhibited his work in London, at the International Design Biennial in St Etienne and have recently become a winner of the ADC Young Guns award in New York.

Koshi Kawachi

Koshi Kawachi was born in 1973 in Gifu, Japan. Graduated from Experimental Design Course, Nagoya University of Arts. He currently lives and works in Tokyo, Japan.

L

Louie Rigano

Louie Rigano is an industrial designer based in the New York metropolitan area. A graduate of the Rhode Island School of Design (RISD) with a degree in industrial design and environmental studies, he also spent a year in Japan as a Fulbright Fellow researching traditional Japanese design philosophy and aesthetics, and their role in modern design and culture. He believes that reason must always reaffirm good form, and maintaining a strong connection to the function and production process of an object creates meaningful aesthetics. His work deals with pared down ideals. He seeks a thoughtful and occasionally humorous meditation on contemporary notions of luxury, utility, and cultural values, conveyed by the most direct and incisive means. His search for moments of unexpected accord between the spectacular and the practical is ongoing.

LZF

The company was founded by Marivi Calvo and Sandro Tothill in 1994. Art Director Marivi Calvo leads a team of freelance designers who make up the LZF family. Burkhard Dämmer, Victor Carrasco, Oskar Cerezo, Miguel Herranz, Ray Power, Bang Design, and Luis Eslava Studio all collaborate on a regular basis with LZF. Calvo also coordinates a team of graphic designers, both inside and outside of the company; in Barcelona Lekuonastudio coordinates image strategy with Calvo using resources such as Inocuo the sign, as well as Lo Siento design studios to add spice where needed. Tothill keeps the machine greased and running, coordinating the commercial and production side with the design side. He also coordinates the LZF blog as well as the amateur photographer and music where included.

M

MAMEMIMOMU

When analyzing the structure of the washbasins in the places where children frequent: schools, leisure centres, parks, playgrounds, clinics, and of course the home, we realized that the normal washbasin adaptations are strictly variations of essential aspects (height, mechanism), without entering into details like water-saving mechanism, ergonomics, visual attraction or security. With our proposal, we attempt to cover the later points.

Måns Salomonsen

Måns Salomonsen is a hands-on person full of ideas. With his patient and hard working personality he's the person who brings every project until the very end. Måns's mature, open and curious attitude for design gives his works a specific character and after all combining that to his handcraft skills and knowledge, nothing is impossible.

Michiel Cornelissen

Michiel Cornelissen (1969) is a Dutch product designer. His designs usually have a functional, innovative background, but they tend to have a strong emotional side as well: a surprise, a touch of humor or a nostalgic connotation.

Michiel graduated with honors in Industrial Design Engineering from Delft University of Technology, the Netherlands. He worked for over a decade in design, creative direction and strategic functions for Philips Design - the prestigious design studio of Dutch giant Philips Electronics - where he contributed to projects ranging from hospital equipment to children's products. The work he has done with various teams at Philips Design has been recognized with awards including IF, ID and Red Dot.

Michiel's own design studio was founded in 2010. Though still working on a wide range of products for various clients, he has also created his own collection of 3d-printed personal accessories and interior objects.

His recent work, including 36 pencil bowl, Pod à porter and PadFoot has been published in various magazines as well as in internet publications such as Gizmodo, Wired, Core77, Mocoloco and FastCoDesign, and many others.

Michiel lives and works in Utrecht, The Netherlands.

Miriam Aust

The German furniture designer, Miriam Aust works and lives in Kassel. She is interested in the way we interact with objects in our daily life and the changes that happen through new developments in society. The human desire for individuality is a topic in her work that reappears in various ways.

The individual product produced in a series is a reflection on this idea. This results in using existing objects, forcing the user to take an active role, or manipulating the production process.

The resulting projects are bookshelves that are constructed around an existing old book through the use of a computer program or furniture that includes plants turning the user into a gardener and designer at the same time.

Since 2011 she shares a studio with her partner Sebastian Amelung who complements her in her work. In their current project 'like paper' they take an active role in the production process making every object unique while maintaining the character of a series.

Monocomplex

Monocomplex is a design studio started up in Janually 2011 by south korean designers Jangwon Cho, Eunsang Hwang, Hyunwoo Park and Taemin Kim. Monocomplex is composed of four people with each different personality with the purpose of joyful communication. It intends to perform a design play containing "simple" philosophy and sincerity in the world mixed up with ambiguity, complexity, and diversity.

They study and create new ideas in the realms of product design, furniture and lighting. Their collaborative efforts have created some of the most interesting modern pieces. In the recent times, Monocomplex is in the spotlight as a one of up-and-coming korean design group.

N

NOSIGNER

NOSIGNER is a design and innovation firm that can re-envision your brand by creating essential experiences. They work within the fields of products, graphics, art, communication, space, and architecture. They believe design thinking can solve problems. They work with sustainability, education, isolated locality. Disaster restoration and break stereotypes by creating fresh design experiences.

O

Olga Kalugina

Olga Kalugina was born in 1985 in Yekaterinburg, Russia. She was graduated in Industrial Design Department at Ural State Academy's of Architecture and Art.

Olga was awarded in many competitions including Russian Design (2008), International Design Award (2009), international competition of young designers KREATA, Design and Design International Award (2011), Be Open Award (2012) and others.

She currently lives and works in Moscow, Russia.

Oriental Fusion Handcraft[SM] - Recycling Workshop

The "Oriental Fusion Handcraft[SM] – Recycling Workshop", is an educational and ecological activity that is promoted by Reiyukai America. Reiyukai America is a non-profit membership-based organization founded in October 1972 in Los Angeles, CA. Ever since its foundation, Reiyukai America has been supporting its members' activities to benefit their local communities.

Reiyukai America has been promoting the development of one's self for the benefit of all humanity through various activities such as "Letter to My Parents Contest[SM]" "Reiyukai Family Day Festival" "Remembering National Heroes" and others.

The "Oriental Fusion Handcraft[SM] – Recycling Workshop" is an activity that started more than 10 years ago in Mexico after searching for the possibility to support the people and its community by providing a method to recycle used materials, mainly milk cartons, into a useful crafts. The activity was introduced to the United States in 2005.

Through this workshop, Reiyukai America seeks to encourage people to share "the importance of transformation and development: the idea of change" Milk cartons are usually looked at as trash. However, once changing the point of view, their value can be seen as something to turn into a beautiful craft. In Reiyukai, it is believed that transformation can be applied in our lives and to ourselves as well.

P

Pavel Eekra

Pavel Eekra is trying to figure out how to construct a space, rediscovering methods and algorithms for creating a whole out of parts through research and experimentation with modern design tools, to get the most of the material, reducing, in this case, the complexity of industrial production. Thus making the design more accessible and harmonious with the environment.

Sources of inspiration:
Spatial geometry, topology, crystallography
The structure of plants, bacteria, radiolarians, corals and other inhabitants of the forests and marine life.
Modern architecture, arts and crafts and travel.

participant DMY Berlin 2011
participant Milan 2011 «Giulio Cappellini present 02.2.20 project» via tortona, 31
winner of the Moscow Design Week 2010
participant Sretenka Design Week 2010

Pavel Sidorenko

Pavel Sidorenko - Estonian designer, was born in 1980 in Tallinn. Studied Product design at Estonian Academy of Arts and Graduated in 2006. Among the awards: Estonia, Tallinn. "Eesti Disain ja Elukeskkond 2009" 3rd prize in 2009; Germany, Frankfurt. AMBIENTE 2010. part of EDL stand and Estonia, Tallinn. "BRUNO" nominate in 2010. His aim is to create functional and playful products that retains its simplicity, interacting with the space and user. To create not only incorporate pragmatic necessity, but also transmit an emotional quality within the everyday environment.

Peleg Design

The world of Peleg Design revolves around three basic principles: functionality, aesthetics, and humor. It is a world motivated by continuous curiosity with the explicit purpose to transform what was once an average, been-there-done-that concept into an innovative and truly exciting experience.

Peleg Design studio was founded in 2004 by Shahar Peleg, who holds a B.A. in Design and Interior Architecture from the Holon Institute of Technology (HIT).

The studio's products are designed to be used in homes, offices and other living spaces. Beyond the end-product, Shahar provides the observer with a complete, all inclusive experience. Through optical illusions that stretch the limits and break conventions, he invites the viewer to take a closer look, revealing a fresh view of everyday products. The viewer is encouraged to discover that beyond the products' functionality there is a surprising game of enchantment and humor to be seen.

The products' design follows a clean and minimalist line. Using a variety of materials and methods, some of the products are mass produced, others are manufactured in limited series while a select few are single, hand-made.

R

Ronen Kadushin

Ronen Kadushin (b. 1964) is an Israeli designer and design educator living in Berlin since 2005. He taught furniture design and design creativity courses at leading Israeli and European design academies since 1993. In 2004 Kadushin developed the Open Design concept, where the designs of his products can be downloaded, copied, modified and produced, much as in Open Source software. Based on this concept, he formed Open Design, a design and production company for furniture, lighting and accessories which are produced in Berlin and sold in Europe and the USA. Open Design products are regularly presented in solo and group exhibitions world wide, and published in professional literature. Kadushin is teaching Open design in universities and speaks at conferences. He is the author of the Open Design Manifesto.

S

Shahaf Ben Abu

Industrial designer Shahaf Ben Abu was born in (1981) sderot, israel. Graduated at "Hadassah College" in Jerusalem, BA Inclusive Industrial Design. Works and lives in Jerusalem Israel. Trying to explore the encounter between Handcraft and industrial design by combining a wide variety of materials, shapes and methods that transmit through the products values of worm, nostalgic and Unique appearance & use. I want that whoever sees my products will say "why the hell nobody thought of this before?" Words of inspiration: Minimalism, Simplicity. Source of inspiration: Nature.

Spell

In 2007 Spell established itself between the banks of the river Rhine and the foothills off the Veluwe forest in the Dutch town of Arnhem.

Spell aspires to create a new dimension of lifestyle products, constantly looking for new ways to engage the public with inspiring designs. To achieve this, Spell cooperates with various passionate and talented designers.

Purposefully keeping fabrication in the Netherlands to ensure the highest possible quality, remaining true to our original designs.

Spell desires to create more space for people to escape their daily routine.

Aiming to create products that are spontaneous by nature, emphasizing freedom and pleasure in use. It is not our goal to make user friendly design alone, we intend to visualize a playful environment to allow users to express their imagination.

Studio Note

I studied an architecture and product design in Japan After that I went to Design Academy Eindhoven in Netherlands.

I went to studio Richard Hutten and MVRDV at the internship.

I set up "studio note" in Japan after graduating from Design Academy.

Study O Protable

Study O Portable is a designing practice for accessories and other portable objects based in London, founded by Bernadette Deddens (the Netherlands) and Tetsuo Mukai (Japan) in April 2009. Their work evolves around the idea of portability and question of functionality in objects. They take great interest in the methods of designing just as much as the resulting work and often choose a simple but considered production process and material.

They met while studying at ArtEz in Arnhem, the Netherlands. Bernadette graduated from Design Products Royal College of Art in 2006 and has since worked as an independent accessories designer and consultant. She is a visiting lecturer at London College of Fashion. She looks for practical emotional functions in objects that we carry around. Tetsuo has a background in fashion design and worked in London and Paris. He investigates alternative ways to design objects.

Studio Veríssimo

Studio Verissimo arises through the collaboration of two designers (Claudio Cardoso and Telma Verissimo).

Since then he has developed projects for Details (Germany), Benetton (Italy) Lungomare Gallery (Italy), Gallery Yron (Lisbon), Appel Design Gallery (Berlin) and the Touch (USA). His works exhibited in many cities, Tokyo, New York, Olhão, Saint Etienne, Lisbon, Miami, Amsterdam, Milan and London.

Projects Studio Verissimo seeks to make people happier through design. The function and mood are the key ingredients to its portfolio of fun and contemporary, that brings together objects lighting, kitchen accessories and eco-design projects resulting from the reuse or recycling of various materials.

The references that inspire the Studio Verissimo projects are varied, from traditional crafts Portuguese, ecology, the changing lifestyles of globalization, what drives the constant interaction between the micro and macro areas of contemporary design are features that Telma and Claudio integrate in a unique and dynamic in the creation of their design projects.

T

Twentyone Studio

Twentyone Studio was established in Rome in 2008 by a group of architects and designers. Their prior experience in the planning and design of homes, yachts, hotels and public establishments allowed these professionals to accumulate a level of knowledge that, when coupled with the technical know-how of specialized companies, translates into quality, high-level design and products. Twentyone's philosophy is based on a contemporary interpretation of history and artisanship; artisanal tradition is applied to new millennium thinking, trying to combine manufacturing skills with technology, scientific research and communication. All this gives rise to a unique mix of tangible and intangible, matter and thought.

V

Verena Klette

1977 born in Stralsund/GER;

1995-1999 apperenticeship as goldsmith in Bergen/ GER;

2000-2004 Hochschule für Getsaltung Pforzheim/ GER, jewellery department since 2007 own studio at Stralsund/GER.

Y

Yong Joo Kim

A Native of Seoul, Korea, Yong Joo Kim is a Niche award winning artist who explores the concept of inner-beauty by investigating the relationship between the process of exploration and trust endowed to a material by an artist.

After graduating from the Rhode Island School of Design with an MFA in Jewelry and Metalsmithing, Yong Joo received the Adrianna Farrelli Prize for Excellence in Fiber Art from the Philadelphia Museum of Art as well as the prestigious Niche award in Jewelry: Sculpture to Wear category. She has also been the recipient of the Professional Arts Development (PAD) grant from the Rhode Island State Council on the Arts (RISCA) and a finalist for the Lydon Emerging Artist Award (LEAP).

Her work has been internationally exhibited at museums and galleries in New York, Boston, Chicago, Miami, Philadelphia, Seattle, Sydney, Firenze and Seoul. Her work has also been selected as part of the prestigious Museum of Arts and Design's (MAD) permanent collection.

Yoshio Hasegawa

Yoshio Hasegawa is a self-taught paper artist lives in Nagoya City in Japan. He realized that creating the art works was the best way to know who he is, when he was fifteen years old and determined to be an artist. While working as a faculty for the facility for the disables, he has been continuing to create the art work made by paper more than twenty years. Hasegawa received numerous awards which includes the grand prize of KAMIWAZA competition, the second prize of Kankaku museum in Japan as well as the honorable mention of Amateras mini paper art exhibition in Sofia. Hasegawa is currently working as the full-time artist.

Most of his pieces are constructed by sliced corrugated paper and illustration board painted with acrylic. The sliced small pieces are connected with glue.

One of his signature pieces, the empty cans represent the objects which already finished their purpose of existence. Hasegawa believes that the sliced corrugated paper is the great material to show his concept, because the airy spaces on the surfaces can emphasize to express the ambiguity of the existence. However, at the same time, he always tries to make the pieces as if they don't exist. The color of white represents his indication. It sounds like a Zen riddle, but he would like to remind the people about the sense of existence by the objects which don't have a strong presence.

Yuruliku

Yuruliku is a design studio, by designers Koushi Ikegami and Kinue Oneda.

Yuruliku was named by combining the words "yururi", which means "relaxing", with "yukkuri", which means "slowly", in Japanese.

We are in search of new styles in stationery that can make everyone feel happy. yuruliku's works are based on a theme "little humor" found in everyday life. We cherish the importance to express "enjoyment" "simplicity" and "fresh discovery" in our works.

Z

Zim & Zou

Lucie Thomas teamed up with Thibault Zimmermann to form Zim & Zou, a french studio based in Nancy that explores different fields including paper sculpture, installation, graphic design, illustration. Both aged 25, they studied graphic design during 3 years in an artschool. Rather than composing images on a computer, they prefer creating real objects with paper and taking photos out of them. A number of intricate illustrations actually come from the three-dimensional installations made by Zim & Zou. Their choice of paper is due to the versatility and good quality of the material, especially when it is sculpted and photographed. Zim & Zou's strength is to be a complementary and polyvalent duo.

Add: Room C, 9/F., Sun House, 181 Des Voeux Road Central, Hong Kong, China
Tel: 852-31840676, Fax: 852-25432396

Add: 13B, Zijing Ge, Fa Zhan Xing Yuan, No.1, Mintian Road, Futian District, Shenzhen, China
Tel: 86-755-82913355, Fax: 86-755-82020029

Web: www.artpower.com.cn
Weibo: @ 艺力出版 --- ACS 杂志
E-mail: artpower@artpower.com.cn

! NEW BOOKS PUBLISHED Artpower 2013 UPCOMING 01 04

TOP Space & Art III
978-988-15743-4-3

215×285 mm
328 pages
Hardback

Public Construction
978-988-15742-2-0

247×330 mm
416 pages
Hardback

The Art of Financial Design
978-7-5611-6827-1

245×290 mm
376 pages
Hardback

Car Parks

245×290 mm
384 pages
Hardback

Old Space & New Idea
978-988-16687-9-0

245×290 mm
296 pages
Hardback

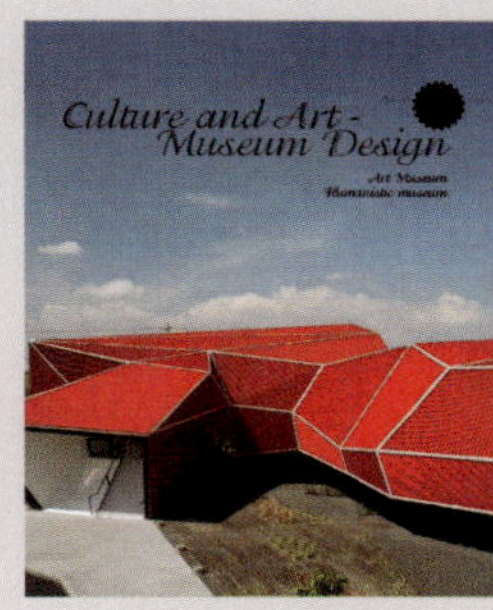

Culture and Art - Museum Design
978-988-19226-7-0

245×280 mm
376 pages
Hardback

Hospital Design
978-988-16428-7-5

215×285 mm
416 pages
Hardback

Scenery & Family Life - Housing Landscape

245×330 mm
376 pages
Hardback

Light in Space Design

245×330 mm
332 pages
Hardback

SPA-DE 17
978-4-89737-713-1

230×283 mm
168 pages
Hardback

Streets and Squares
978-988-16428-9-1

245×330 mm
456 pages
Hardback

European Style Villa Design
978-988-15743-0-5

245×290 mm
344 pages
Hardback

Resorts & SPA

245×290 mm
352 pages
Hardback

Adorn Your Life
978-988-15743-8-1

245×290 mm
312 pages
Hardback

! NEW ARRIVALS

房地产企业高效运营管理全案（上下册）
978-7-5611-7283-4

210×285 mm
738 pages
Hardback

破解商业地产定位密码
978-7-80653-884-5

210×285 mm
532 pages
Hardback

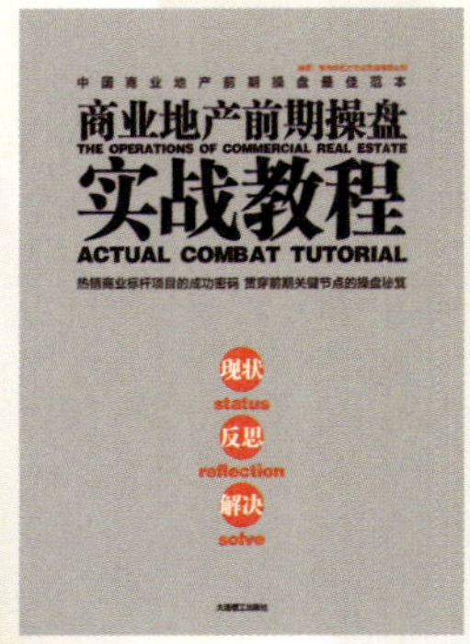

商业地产前期操盘实战教程

210×285 mm
437 pages
Hardback

Pack Your Life II
978-988-15743-7-4

215×285 mm
360 pages
Hardback

DIY
978-988-15743-6-7

210×270 mm
264 pages
Hardback

Touch Branding III
978-988-16428-1-7

215×285 mm
320 pages
Hardback

Design Art of Villa II

245×300 mm
408 pages
Hardback

SPA-DE 18
978-4-89737-731-5

230×285 mm
172 pages
Hardback

Material+: Creative Products
978-988-16687-5-2

215×285 mm
272 pages
Hardback

TOP Villas Of International Style
978-988-15743-9-8

310×450 mm
479 pages
Hardback

First Taste Hotel! - Global Hotel Design
978-988-19735-0-4

310×450 mm
592 pages
Hardback

Cheers! Wine Cellar Design
978-988-15743-5-0

245×290 mm
296 pages
Hardback

Dessert Station
978-988-15743-2-9

245×290 mm
360 pages
Hardback

Stylish Stores
978-988-19735-9-7

245 X 300 mm
232 pages+240 pages
Hardback

Audiovisual Communication
978-988-15743-3-6

247×330 mm
424 pages
Hardback

Salon.com
978-988-15742-9-9

245×290 mm
248 pages
Hardback

Walk and Watch - Display Design
978-988-19735-4-1

245×290 mm
320 pages
Hardback

Design Art of Villa
978-988-15743-1-2

245×290 mm
360 pages
Hardback

Restaurant & Bar Design II
978-988-15742-5-1

245×300 mm
344 pages
Hardback

The World SPA Design
978-988-19735-8-0

245 X 290 mm
248 pages+264 pages
Hardback

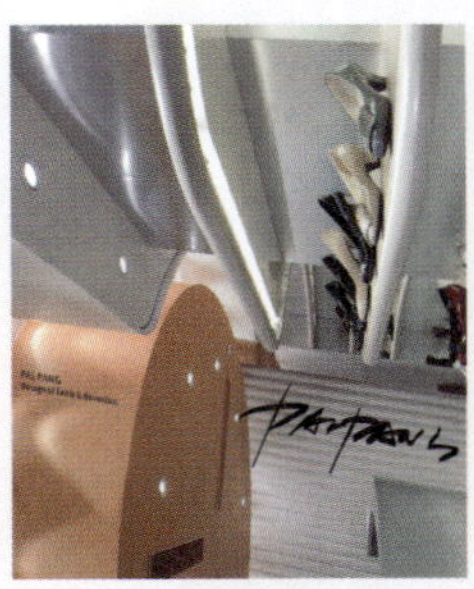

Pal Pang - Design of Faith and Devotion
978-7-5611-5595-0

185×260 mm
279 pages
Hardback

inSIDE/outSIDE Office Design II
978-988-15742-6-8

247×330 mm
390 pages
Hardback

inSIDE/outSIDE Office Design
978-988-18893-8-6

240×295 mm
485 pages
Hardback

Design Feast
978-988-68578-2-3

220×290 mm
287 pages
Hardback

Living Space
978-988-15742-1-3

247×330 mm
368 pages
Hardback

Fashion Sample House Design
978-7-5611-6308-5

245×290 mm
311 pages
Hardback

TOP Space & Art II
978-4-568-50432-3

220X290 mm
280 pages
Hardback

TOP Space & Art
978-988-17684-6-9

210×270 mm
255 pages
Hardback

More ? Mall !
978-988-18893-6-2

285×285 mm
328 pages
Hardback

Quality Life - Top Villa Design
978-7-5611-6518-8

245×290 mm
352 pages
Hardback

Sakura Villa
978-988-68578-5-4

280×280 mm
344 pages
Hardback

SPA-DE 15
978-4-8973-7680-6

230×285 mm
168 pages
Hardback

SPA-DE 16
978-4-8973-7698-1

210×285 mm
168 pages
Hardback

Restaurants & Bars Design
978-988-18893-4-8

245×270 mm
416 pages
Hardback

New Sales Office Design
978-7-5611-6228-6

220×280 mm
344 pages
Hardback

Commercial Space
978-988-18893-2-4

235×305 mm
352 pages
Hardback

inSIDE outSIDE Residence
978-4-903233-68-0

245×305 mm
432 pages
Hardback

Furniture Exhibition Design
978-4-903233-54-3

220×300 mm
408 pages
Hardback

New Typography

978-988-15742-8-2

210×270 mm
280 pages
Hardback

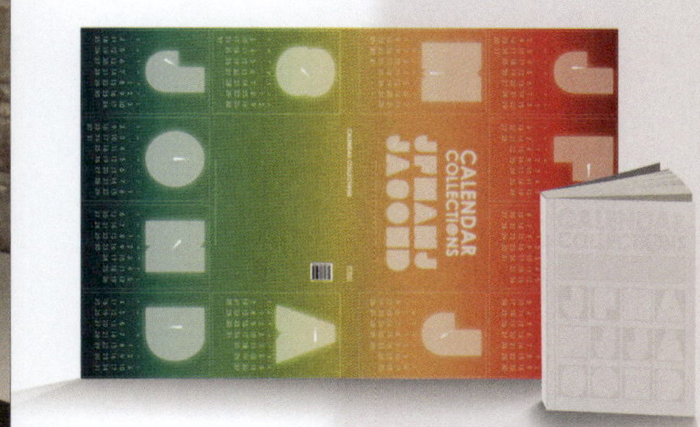

Calendar Collections

978-4-568-50443-9

220×280 mm
256 pages
Paperback

359° Image
- Hong Zhongxuan Global Journey Design Elements

978-7-5611-6613-0

230×300 mm
360 pages
Paperback

Paper Art

978-988-15742-0-6

210×270 mm
264+232 pages
Hardback

Way of the Sign II

978-988-19735-6-6

210×290 mm
280 pages
Hardback

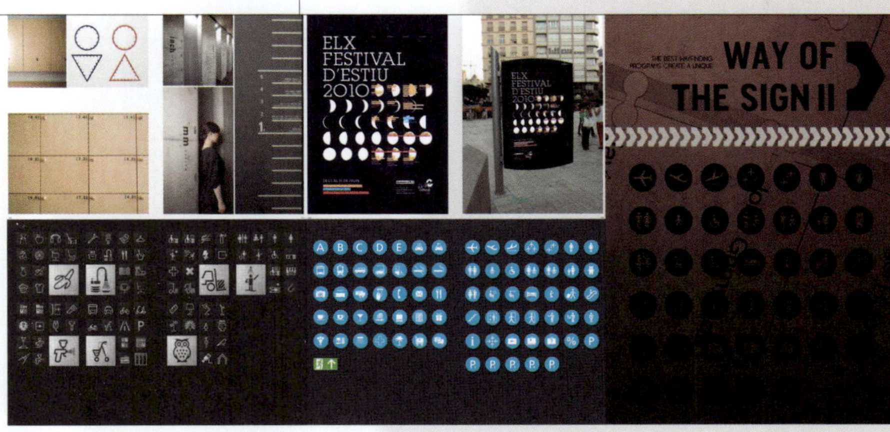

Pack Your Life
978-4-903233-72-7

200×270 mm
326 pages
Hardback

Touch Branding II
978-988-19125-3-4

210×270 mm
264 pages
Hardback

Type Image
978-7-5611-6002-2

185×260 mm
224 pages
Hardback

Sales Promotion Design
978-4-568-50455-2

225×297mm
288 pages
hardback

Way of the Sign
978-4-903233-66-6

210×270 mm
264 pages
Hardback

Logos Talk
978-1-58423-444-9

185×260 mm
264 pages
Hardback

Fabulous Furniture
978-988-19735-5-9

245×290 mm
272 pages
Hardback

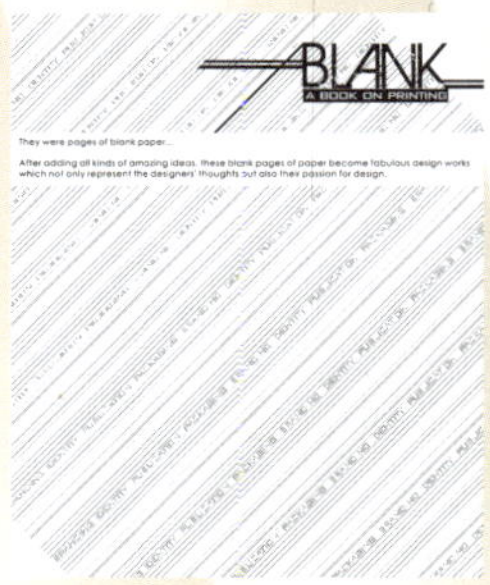

≠ BLANK
- A Book on Printing
978-988-18893-5-5

200×280 mm
320 pages
Paperback

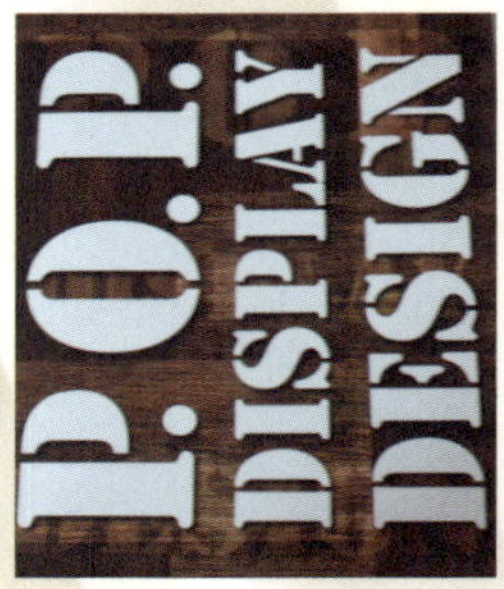

P.O.P. Display Design
978-988-18893-7-9

225×297 mm
256 pages
Hardback

Graphic Moment
978-4-903233-71-0

210×270 mm
304 pages
Hardback

Graphic×Space
978-4903-2336-5-9

220×280 mm
240 pages
Hardback

WWW & Design
978-988-99550-5-2

200×210 mm
Hardback
357 pages

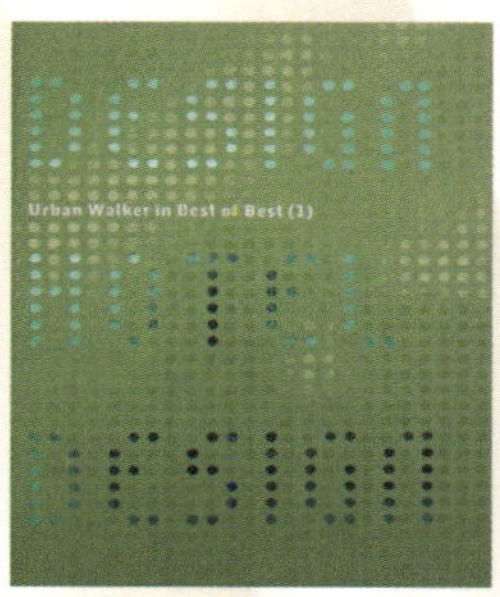

Design Hotel Design
978-988-17684-3-8

210×285 mm
236 pages(vol.1)
223 pages(vol.2)
Hardback

The Image of Business Card Today
978-988-17684-7-6

200×260 mm
320 pages
Hardback

Fashion Hangtag Design
978-988-18893-3-1

210×270 mm
285 pages
Hardback

ARTPOWER

Acknowledgement

We would like to thank all the designers and companies who made significant contributions to the compilation of this book. Without them, this project would not have been possible. We would also like to thank many others whose names did not appear on the credits, but made specific input and support for the project from beginning to end.

Future Editions

If you would like to contribute to the next edition of Artpower, please email us your details to: artpower@artpower.com.cn